Introduction to Python for Science and Engineering

Series in Computational Physics

Parallel Science and Engineering Applications: The Charm++ Approach
Laxmikant V. Kale, Abhinav Bhatele

Introduction to Numerical Programming: A Practical Guide for Scientists and Engineers Using Python and C/C++
Titus A. Beu

Computational Problems for Physics: With Guided Solutions Using Python
Rubin H. Landau, Manual José Páez

Introduction to Python for Science and Engineering
David J. Pine

For more information about this series, please visit: https://www.crcpress.com/Series-in-Computational-Physics/book-series/CRCSERCOMPHY

Introduction to Python for Science and Engineering

David J. Pine

CRC Press
Taylor & Francis Group
Boca Raton London New York

CRC Press is an imprint of the
Taylor & Francis Group, an **informa** business

MATLAB® is a trademark of The MathWorks, Inc. and is used with permission. The MathWorks does not warrant the accuracy of the text or exercises in this book. This book's use or discussion of MATLAB® software or related products does not constitute endorsement or sponsorship by The MathWorks of a particular pedagogical approach or particular use of the MATLAB® software.

CRC Press
Taylor & Francis Group
6000 Broken Sound Parkway NW, Suite 300
Boca Raton, FL 33487-2742

© 2019 by Taylor & Francis Group, LLC

CRC Press is an imprint of Taylor & Francis Group, an Informa business

No claim to original U.S. Government works

Printed on acid-free paper

International Standard Book Number-13: 978-1-138-58389-4 (Paperback)
International Standard Book Number-13: 978-1-138-58390-0 (Hardback)

This book contains information obtained from authentic and highly regarded sources. Reasonable efforts have been made to publish reliable data and information, but the author and publisher cannot assume responsibility for the validity of all materials or the consequences of their use. The authors and publishers have attempted to trace the copyright holders of all material reproduced in this publication and apologize to copyright holders if permission to publish in this form has not been obtained. If any copyright material has not been acknowledged please write and let us know so we may rectify in any future reprint.

Except as permitted under U.S. Copyright Law, no part of this book may be reprinted, reproduced, transmitted, or utilized in any form by any electronic, mechanical, or other means, now known or hereafter invented, including photocopying, microfilming, and recording, or in any information storage or retrieval system, without written permission from the publishers.

For permission to photocopy or use material electronically from this work, please access www.copyright.com (http://www.copyright.com/) or contact the Copyright Clearance Center, Inc. (CCC), 222 Rosewood Drive, Danvers, MA 01923, 978-750-8400. CCC is a not-for-profit organization that provides licenses and registration for a variety of users. For organizations that have been granted a photocopy license by the CCC, a separate system of payment has been arranged.

Trademark Notice: Product or corporate names may be trademarks or registered trademarks, and are used only for identification and explanation without intent to infringe.

Library of Congress Cataloging-in-Publication Data
Names: Pine, David J., author. Title: Introduction to Python for science and engineering / by David J. Pine. Description: Boca Raton, Florida : CRC Press, [2019] | Series: Series in computational physics | Includes bibliographical references and index. Identifiers: LCCN 2018027880 (print) | LCCN 2018051956 (ebook) | ISBN 9780429506413 (eBook General) | ISBN 9780429014253 (eBook Adobe Reader) | ISBN 9780429014246 (eBook ePub) | ISBN 9780429014239 (eBook Mobipocket) | ISBN 9781138583894 (paperback : acid-free paper) | ISBN 9781138583900 (hardback : acid-free paper). Subjects: LCSH: Python (Computer program language) | Computer programming. | Engineering--Data processing. | Science--Data processing. Classification: LCC QA76.73.P98 (ebook) | LCC QA76.73.P98 P58 2019 (print) | DDC 005.13/3--dc23 LC record available at https://lccn.loc.gov/2018027880

Visit the Taylor & Francis Web site at
http://www.taylorandfrancis.com

and the CRC Press Web site at
http://www.crcpress.com

To Alex Pine
who introduced me to Python

Contents

Preface   *xv*

About the Author   *xix*

1 Introduction   1
   1.1 Introduction to Python for Science and Engineering .   1

2 Launching Python   3
   2.1 Interacting with Python   3
   2.2 Installing Python on Your Computer   4
   2.3 The Spyder Window   4
   2.4 The IPython Pane   4
       2.4.1 Magic commands   6
       2.4.2 System shell commands   8
       2.4.3 Tab completion   8
       2.4.4 Recap of commands   9
   2.5 Interactive Python as a Calculator   9
       2.5.1 Binary arithmetic operations in Python   10
       2.5.2 Types of numbers   10
       2.5.3 Important note on integer division in Python .   12
   2.6 Variables   13
       2.6.1 Names and the assignment operator   13
       2.6.2 Legal and recommended variable names   14
       2.6.3 Reserved words in Python   15
   2.7 Script Files and Programs   16
       2.7.1 First scripting example: The Editor pane   16
   2.8 Python Modules   18
       2.8.1 Python modules and functions: A first look . .   20
       2.8.2 Some NumPy functions   22
       2.8.3 Scripting Example 2   23
       2.8.4 Different ways of importing modules   24
   2.9 Getting Help: Documentation in IPython   26

		2.10 Stand-alone IPython	26
		2.10.1 Writing Python scripts in a text editor	27
	2.11	Programming Errors	28
		2.11.1 Pyflakes	28
		2.11.2 Error checking	29
	2.12	Exercises	29
3	Strings, Lists, Arrays, and Dictionaries		33
	3.1	Strings	34
	3.2	Lists	35
		3.2.1 Slicing lists	37
		3.2.2 The `range` function: Sequences of numbers	38
		3.2.3 Tuples	39
		3.2.4 Multidimensional lists and tuples	40
	3.3	NumPy Arrays	41
		3.3.1 Creating arrays (1-d)	41
		3.3.2 Mathematical operations with arrays	43
		3.3.3 Slicing and addressing arrays	46
		3.3.4 Fancy indexing: Boolean masks	47
		3.3.5 Multi-dimensional arrays and matrices	49
		3.3.6 Differences between lists and arrays	52
	3.4	Dictionaries	53
	3.5	Objects	55
	3.6	Exercises	57
4	Input and Output		61
	4.1	Keyboard Input	61
	4.2	Screen Output	64
		4.2.1 Formatting output with `str.format()`	64
		4.2.2 Printing arrays	68
	4.3	File Input	69
		4.3.1 Reading data from a text file	69
		4.3.2 Reading data from an Excel file: CSV files	71
	4.4	File Output	73
		4.4.1 Writing data to a text file	73
		4.4.2 Writing data to a CSV file	76
	4.5	Exercises	76

Contents ix

5	**Conditionals and Loops**	**81**
5.1	Conditionals	82
	5.1.1 `if`, `elif`, and `else` statements	82
	5.1.2 Logical operators	86
5.2	Loops	87
	5.2.1 `for` loops	87
	5.2.2 `while` loops	91
	5.2.3 Loops and array operations	93
5.3	List Comprehensions	94
5.4	Exercises	96
6	**Plotting**	**99**
6.1	An Interactive Session with PyPlot	100
6.2	Basic Plotting	102
	6.2.1 Specifying line and symbol types and colors	106
	6.2.2 Error bars	108
	6.2.3 Setting plotting limits and excluding data	110
	6.2.4 Subplots	113
6.3	Logarithmic Plots	116
	6.3.1 Semi-log plots	116
	6.3.2 Log-log plots	118
6.4	More Advanced Graphical Output	118
	6.4.1 An alternative syntax for a grid of plots	122
6.5	Plots with multiple axes	125
6.6	Mathematics and Greek symbols	126
6.7	The Structure of matplotlib: OOP and All That	131
	6.7.1 The backend layer	132
	6.7.2 The artist layer	135
	6.7.3 The PyPlot (scripting) layer	137
6.8	Contour and Vector Field Plots	139
	6.8.1 Making a 2D grid of points	139
	6.8.2 Contour plots	140
	6.8.3 Streamline plots	144
6.9	Three-Dimensional Plots	149
6.10	Exercises	152
7	**Functions**	**155**
7.1	User-Defined Functions	156
	7.1.1 Looping over arrays in user-defined functions	158

	7.1.2	Fast array processing for user-defined functions	160
	7.1.3	Functions with more than one input or output	161
	7.1.4	Positional and keyword arguments	162
	7.1.5	Variable number of arguments	163
	7.1.6	Passing function names and parameters as arguments	164
7.2	Passing data (objects) to and from functions	167
	7.2.1	Variables and arrays created entirely within a function	167
	7.2.2	Passing lists and arrays to functions: Mutable and immutable objects	169
7.3	Anonymous Functions: `lambda` Expressions	171
7.4	NumPy Object Attributes: Methods and Instance Variables	173
7.5	Example: Linear Least Squares Fitting	175
	7.5.1	Linear regression	177
	7.5.2	Linear regression with weighting: χ^2	179
7.6	Exercises	182

8 Curve Fitting — 187

8.1	Using Linear Regression for Fitting Nonlinear Functions	187
	8.1.1 Linear regression for fitting an exponential function	187
	8.1.2 Linear regression for fitting a power-law function	192
8.2	Nonlinear Fitting	193
8.3	Exercises	198

9 Numerical Routines: SciPy and NumPy — 205

9.1	Special Functions	206
9.2	Random Numbers	209
	9.2.1 Uniformly distributed random numbers	210
	9.2.2 Normally distributed random numbers	210
	9.2.3 Random distribution of integers	211
9.3	Linear Algebra	212
	9.3.1 Basic computations in linear algebra	212
	9.3.2 Solving systems of linear equations	213
	9.3.3 Eigenvalue problems	214

Contents

- 9.4 Solving Nonlinear Equations ... 216
  - 9.4.1 Single equations of a single variable ... 217
  - 9.4.2 Solving systems of nonlinear equations ... 221
- 9.5 Numerical Integration ... 221
  - 9.5.1 Single integrals ... 222
  - 9.5.2 Double integrals ... 226
- 9.6 Solving ODEs ... 227
- 9.7 Discrete (Fast) Fourier Transforms ... 231
  - 9.7.1 Continuous and discrete Fourier transforms ... 231
  - 9.7.2 The SciPy FFT library ... 232
- 9.8 Exercises ... 234

10 Data Manipulation and Analysis: Pandas ... 239
- 10.1 Reading Data from Files Using Pandas ... 240
  - 10.1.1 Reading from Excel files saved as csv files ... 240
  - 10.1.2 Reading from text files ... 247
  - 10.1.3 Reading from an Excel file ... 250
- 10.2 Dates and Times in Pandas ... 251
- 10.3 Data Structures: Series and DataFrame ... 253
  - 10.3.1 Series ... 253
  - 10.3.2 DataFrame ... 256
- 10.4 Getting Data from the Web ... 261
- 10.5 Extracting Information from a DataFrame ... 263
- 10.6 Plotting with Pandas ... 267
- 10.7 Grouping and Aggregation ... 272
  - 10.7.1 The `groupby` method ... 273
  - 10.7.2 Iterating over groups ... 274
  - 10.7.3 Reformatting DataFrames ... 277
  - 10.7.4 Custom aggregation of DataFrames ... 280
- 10.8 Exercises ... 281

11 Animation ... 287
- 11.1 Animating a Sequence of Images ... 287
  - 11.1.1 Simple image sequence ... 288
  - 11.1.2 Annotating and embellishing videos ... 292
- 11.2 Animating Functions ... 294
  - 11.2.1 Animating for a fixed number of frames ... 295
  - 11.2.2 Animating until a condition is met ... 300
- 11.3 Combining Videos with Animated Functions ... 306

	11.3.1 Using a single animation instance	307
	11.3.2 Combining multiple animation instances	308
11.4	Exercises	311

12 Python Classes and GUIs — 315

- 12.1 Defining and Using a Class 316
  - 12.1.1 The `__init__()` method 319
  - 12.1.2 Defining methods for a class 320
  - 12.1.3 Calling methods from within a class 321
  - 12.1.4 Updating instance variables 322
- 12.2 Inheritance 323
- 12.3 Graphical User Interfaces (GUIs) 326
  - 12.3.1 Event-driven programming 327
  - 12.3.2 PyQt 328
  - 12.3.3 A basic PyQt dialog 328
  - 12.3.4 Summary of PyQt5 classes used 337
  - 12.3.5 GUI summary 337

A Installing Python — 339

- A.1 Installing Python 339
  - A.1.1 Setting preferences 340
  - A.1.2 Pyflakes 340
  - A.1.3 Updating your Python installation 341
- A.2 Testing Your Installation of Python 341
- A.3 Installing FFmpeg for Saving Animations 343

B Jupyter Notebooks — 345

- B.1 Launching a Jupyter Notebook 345
- B.2 Running Programs in a Jupyter Notebook 347
- B.3 Annotating a Jupyter Notebook 348
  - B.3.1 Adding headings and text 349
  - B.3.2 Comments with mathematical expressions ... 350
- B.4 Terminal commands in a Jupyter notebook 351
- B.5 Plotting in a Jupyter Notebook 351
- B.6 Editing and Rerunning a Notebook 353
- B.7 Quitting a Jupyter Notebook 353
- B.8 Working with an Existing Jupyter Notebook 353

C Glossary — 355

D	Python Resources	359
	D.1 Python Programs and Data Files Introduced in This Text	359
	D.2 Web Resources	359
	D.3 Books	361

Index   *363*

Preface

The aim of this book is to provide science and engineering students a practical introduction to technical programming in Python. It grew out of notes I developed for various undergraduate physics courses I taught at NYU. While it has evolved considerably since I first put pen to paper, it retains its original purpose: to get students with no previous programming experience writing and running Python programs for scientific applications with a minimum of fuss.

The approach is pedagogical and "bottom up," which means starting with examples and extracting more general principles from that experience. This is in contrast to presenting the general principles first and then examples of how those general principles work. In my experience, the latter approach is satisfying only to the instructor. Much computer documentation takes a top-down approach, which is one of the reasons it's frequently difficult to read and understand. On the other hand, once examples have been seen, it's useful to extract the general ideas in order to develop the conceptual framework needed for further applications.

In writing this text, I assume that the reader:

- has never programmed before;
- is not familiar with programming environments;
- is familiar with how to get around a Mac or PC at a very basic level; and
- is competent in basic algebra, and for Chapters 8 and 9, calculus, linear algebra, ordinary differential equations, and Fourier analysis. The other chapters, including 10–12, require only basic algebra skills.

This book introduces, in some depth, four Python packages that are important for scientific applications:

NumPy, short for Numerical Python, provides Python with a multidimensional array object (like a vector or matrix) that is at the center of virtually all fast numerical processing in scientific Python.

It is both versatile and powerful, enabling fast numerical computation that, in some cases, approaches speeds close to those of a compiled language like C, C++, or Fortran.

SciPy, short for Scientific Python, provides access through a Python interface to a very broad spectrum of scientific and numerical software written in C, C++, and Fortran. These include routines to numerically differentiate and integrate functions, solve differential equations, diagonalize matrices, take discrete Fourier transforms, perform least-squares fitting, as well as many other numerical tasks.

matplotlib is a powerful plotting package written for Python and capable of producing publication-quality plots. While there are other Python plotting packages available, matplotlib is the most widely used and is the *de facto* standard.

Pandas is a powerful package for manipulating and analyzing data formatted and labeled in a manner similar to a spreadsheet (think Excel). Pandas is very useful for handling data produced in experiments, and is particularly adept at manipulating large data sets in different ways.

In addition, Chapter 12 provides a brief introduction to Python classes and to PyQt5, which provides Python routines for building graphical user interfaces (GUIs) that work on Macs, PCs, and Linux platforms.

Chapters 1–7 provide the basic introduction to scientific Python and should be read in order. Chapters 8–12 do not depend on each other and, with a few mild caveats, can be read in any order.

As the book's title implies, the text is focused on scientific uses of Python. Many of the topics that are of primary importance to computer scientists, such as object-oriented design, are of secondary importance here. Our focus is on learning how to harness Python's ability to perform scientific computations quickly and efficiently.

The text shows the reader how to interact with Python using IPython, which stands for Interactive Python, through one of three different interfaces, all freely available on the web: Spyder, an integrated development environment, Jupyter Notebooks, and a simple IPython terminal. Chapter 2 provides an overview of Spyder and an introduction to IPython, which is a powerful interactive environment tailored to scientific use of Python. Appendix B provides an introduction to Jupyter notebooks.

Python 3 is used exclusively throughout the text with little reference to any version of Python 2. It's been nearly 10 years since Python 3 was introduced and there is little reason to write new code in Python 2; all the major Python packages have been updated to Python 3. Moreover, once Python 3 has been learned, it's a simple task to learn how Python 2 differs, which may be needed to deal with legacy code. There are many lucid web sites dedicated to this sometimes necessary but otherwise mind-numbing task.

The scripts, programs, and data files introduced in this book are available at https://github.com/djpine/python-scieng-public.

Finally, I would like to thank Étienne Ducrot, Wenhai Zheng, and Stefano Sacanna for providing some of the data and images used in Chapter 11, and Mingxin He and Wenhai Zheng for their critical reading of early versions of the text.

About the Author

David Pine has taught physics and chemical engineering for over 30 years at four different institutions: Cornell University (as a graduate student), Haverford College, UCSB, and, at NYU, where he is a Professor of Physics, Mathematics, and Chemical & Biomolecular Engineering. He has taught a broad spectrum of courses, including numerical methods. He does research in experimental soft-matter physics, which is concerned with materials such as polymers, emulsions, and colloids. These materials constitute most of the material building blocks of biological organisms.

CHAPTER 1

Introduction

1.1 Introduction to Python for Science and Engineering

This book is meant to serve as an introduction to the Python programming language and its use for scientific computing. It's ok if you have never programmed a computer before. This book will teach you how to do it from the ground up.

The Python programming language is useful for all kinds of scientific and engineering tasks. You can use it to analyze and plot data. You can also use it to numerically solve science and engineering problems that are difficult or even impossible to solve analytically.

While we want to marshal Python's powers to address scientific problems, you should know that Python is a general purpose computer language that is widely used to address all kinds of computing tasks, from web applications to processing financial data on Wall Street and various scripting tasks for computer system management. Over the past decade it has been increasingly used by scientists and engineers for numerical computations, graphics, and as a "wrapper" for numerical software originally written in other languages, like Fortran and C.

Python is similar to MATLAB®, another computer language that is frequently used in science and engineering applications. Like MATLAB®, Python is an *interpreted* language, meaning you can run your code without having to go through an extra step of compiling, as required for the C and Fortran programming languages. It is also a *dynamically typed language*, meaning you don't have to declare variables and set aside memory before using them.[1]

Don't worry if you don't know exactly what these terms mean. Their primary significance for you is that you can write Python code, test, and use it quickly with a minimum of fuss.

One advantage of Python compared to MATLAB® is that it is free. It can be downloaded from the web and is available on all the standard computer platforms, including Windows, macOS, and Linux.

---

[1] Appendix C contains a glossary of terms you may find helpful.

This also means that you can use Python without being tethered to the internet, as required for commercial software that is tied to a remote license server.

Another advantage is Python's clean and simple syntax, including its implementation of *object-oriented* programming. This should not be discounted; Python's rich and elegant syntax renders a number of tasks that are difficult or arcane in other languages either simpler or more understandable in Python.

An important disadvantage is that Python programs can be slower than compiled languages like C. For large-scale simulations and other demanding applications, there can be a considerable speed penalty in using Python. In these cases, C, C++, or Fortran is recommended, although intelligent use of Python's array processing tools contained in the NumPy module can greatly speed up Python code. Another disadvantage is that, compared to MATLAB®, Python is less well documented. This stems from the fact that it is public *open source* software and thus is dependent on volunteers from the community of developers and users for documentation. The documentation is freely available on the web but is scattered among a number of different sites and can be terse. This manual will acquaint you with the most commonly used web sites. Search engines like Google can help you find others.

You are not assumed to have had any previous programming experience. However, the purpose of this manual isn't to teach you the principles of computer programming; it's to provide a very practical guide to getting started with Python for scientific computing. Perhaps once you see some of the powerful tasks that you can accomplish with Python, you will be inspired to study computational science and engineering, as well as computer programming, in greater depth.

CHAPTER 2

Launching Python

> *In this chapter you learn about **IPython**, an interface that allows you to use Python interactively with tools that have been optimized for mathematical and computational tasks. You learn how to use IPython as a calculator and how to add, subtract, multiply, divide, and perform other common mathematical functions. You also learn the basic elements of the Python programming language, including **functions**, **variables**, and **scripts**, which are rudimentary computer programs. We introduce Python **modules**, which extend the capabilities of the core Python language and allow you to perform advanced mathematical tasks. You also learn some new ways to navigate your computer's file directories. Finally, you learn how to get help with Python commands and functions.*

2.1 Interacting with Python

There are many different ways to interact with Python. For general purpose use, people typically use the *Python command shell*, which is also called the *Python Interpreter* or *Console*. A shell or console is just a window on your computer that you use to issue written commands. For scientific Python, which is what we are concerned with here, people generally use the *IPython* shell (or console). It has been specifically designed for scientific and engineering use.

There are different ways to launch an IPython shell and write Python code. As a beginner, we recommend using an *Integrated Development Environment* or IDE such as Spyder, a popular IDE that we introduce in the following sections. Spyder uses an IPython shell and provides other features that make it a convenient platform for you to learn about Python. Eventually, you will want to learn about other ways of interacting with Python, such as *Jupyter Notebooks*, which are described in Appendix B. Alternatively, you can interact with Python by writing code using a simple text editor and then running the code from an IPython shell. We describe how to do this in §2.10 towards the end of this chapter. In the end, you should learn to interact with Python in all these ways, as each is valuable, depending on the application. For now, however, we begin our exploration of Python with the Spyder IDE.

2.2 Installing Python on Your Computer

If you haven't already installed Python on your computer, see Appendix A, which includes instructions for installing Python on Macs running under macOSX and on PCs running under Windows.

Once you have installed Python, launch Spyder as directed in Appendix A, and wait for the Spyder window to appear, like the one shown in Fig. 2.1.

2.3 The Spyder Window

The default Spyder window has three panes: the IPython pane, the Editor pane, and the Help pane. The **IPython pane** is the primary way that you interact with Python. You can use it to run Python computer programs, test snippets of Python code, navigate through your computer file directories, and perform system tasks like creating, moving, and deleting files and directories. You will use the **Editor Pane** to write and edit Python programs (or scripts), which are simply sequences of Python commands (code) stored in a file on your computer. The **Help Pane** in Spyder gives help on Python commands.

The individual panes in the Spyder window are reconfigurable and detachable but we will leave them pretty much as they are. However, you may want to adjust the overall size of the window to suit your computer screen. You can find more information about Spyder using the *Help* menu.

2.4 The IPython Pane

The default input prompt of the IPython pane looks like this:

```
In [1]:
```

This prompt signifies that Spyder is running the **IPython** shell. The IPython shell has been specifically designed for scientific and engi-

Launching Python

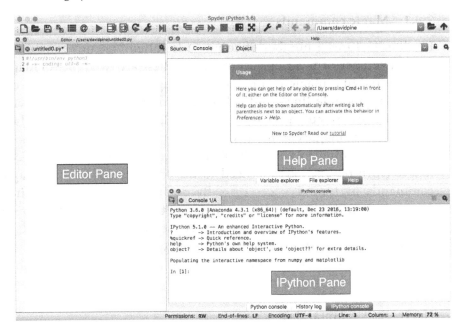

Figure 2.1 Spyder IDE window.

neering use. The standard Python interactive shell uses the prompt >>>. You can pretty much do everything you want to do with either shell, but we will be using the IPython shell as we want to take advantage of some of its special features for scientific computing.

By typing commands at the prompt, IPython can be used to perform various system tasks, such as running programs and creating and moving files around on your computer. This is a different kind of computer interface than the icon-based interface (or graphical user interface, GUI) that you usually use to communicate with your computer. While it may seem more cumbersome for some tasks, it can be more powerful for other tasks, particularly those associated with programming.

Before getting started, we point out that like most modern computer languages, Python is *case sensitive*. That is, Python distinguishes between upper- and lower-case letters. Thus, two words spelled the same but having different letters capitalized are treated as different names in Python. Keep that in mind as we introduce different commands.

2.4.1 Magic commands

IPython features a number of commands called "magic" commands that let you perform various useful tasks. There are two types of magic commands, line magic commands that begin with %—these are executed on a single line—and cell magic commands that begin with %%—these are executed on several lines. Here, we concern ourselves only with line magic commands.

The first thing to know about magic commands is that you can toggle (turn on and off) the need to use the % prefix for line magic commands by typing %automagic. By default, the Automagic switch is set to ON in the Spyder IDE so you don't need the % prefix. To set Automagic to OFF, simply type %automagic at the IPython prompt. Cell magic commands always need the %% prefix.

In what follows below, we assume that Automagic is OFF and thus use the % sign for magic commands.

Navigation commands

IPython recognizes several common navigation commands that are used under the Unix/Linux operating systems. In the IPython shell, these few commands work on Macs, PCs, and Linux machines.

At the IPython prompt, type %cd ~ (*i.e.*, "%cd" – "space" – "tilde", where tilde is found near the upper left corner of most keyboards). This will set your computer to its home (default) directory.

```
In [1]: %cd ~
/Users/pine
```

Next type %pwd (**p**rint **w**orking **d**irectory) and press RETURN. The console should return the path of the current directory of your computer. It might look like this on a Mac:

```
In [2]: %pwd
Out[2]: '/Users/pine'
```

or this on a PC:

```
In [3]: %pwd
Out[3]: C:\\Users\\pine
```

Typing %cd .. ("%cd" – "space" – two periods) moves the IPython shell up one directory in the directory tree, as illustrated by the set of commands below.

```
In [4]: %cd ..
/Users
```

Launching Python

```
In [5]: %pwd
Out[5]: '/Users'
```

The directory moved up one from /Users/pine to /Users. Now type ls (list) and press RETURN. The console should list the names of the files and subdirectories in the current directory.

```
In [6]: %ls
Shared/    pine/
```

In this case, there are only two directories (indicated by the slash) and no files (although the names of the files may be different for you). Type %cd ~ again to return to your home directory and then type pwd to verify where you are in your directory tree.

Making a directory

Let's create a directory within your documents directory that you can use to store your Python programs. We will call it programs. First, return to your home directory by typing %cd ~. Then type %ls to list the files and directories in your home directory.

```
In [7]: %cd ~
/Users/pine

In [8]: %ls
Applications/    Library/    Pictures/
Desktop/         Movies/     Public/
Documents/       Music/
Downloads/       News/
```

To create a directory called programs, type %mkdir programs (**make directory**). Then type %ls to confirm that you have created programs.

```
In [9]: %mkdir programs

In [10]: %ls
Applications/    Library/    Pictures/
Desktop/         Movies/     Public/
Documents/       Music/      programs/
Downloads/       News/
```

You should see that a new directory named programs has been added to the list of directories. Next, type %cd programs to navigate to that new directory.

```
In [11]: %cd programs
/Users/pine/programs
```

Sometimes, the IPython shell becomes cluttered. You can clean up the shell by typing `%clear`, which will give you a fresh shell window.

The `%run` magic command

A very important magic command is `%run` *filename* where *filename* is the name of a Python program you have created. We haven't done this yet but include it here just for reference. We will come back to this later in the chapter.

There are a lot of other magic commands, most of which we don't need, and others that we will introduce as we need them. If you are curious about them, you can get a list of them by typing `%lsmagic`.

2.4.2 System shell commands

You can also run system shell commands from the IPython shell by typing `!` followed by a system shell command. For Macs running OSX and for Linux machines, this means that Unix (or equivalently Linux) commands can be issued from the IPython prompt. For PCs, this means that Windows (DOS) commands can be issued from the IPython prompt. For example, typing `!ls` (**list**) and pressing RETURN lists all the files in the current directory on a Mac. Typing `!dir` on a PC does essentially the same thing (note that system shell commands in Windows are not case sensitive).

2.4.3 Tab completion

IPython also incorporates a number of shortcuts that make using the shell more efficient. One of the most useful is **tab completion**. Let's assume you have been following along and that your are in the directory `Documents` or `My Documents`. To switch to the directory `programs`, you could type `cd programs`. Instead of doing that, type `cd prog` and then press the TAB key. This will complete the command, provided there is no ambiguity in how to finish the command. In the present case, that would mean that there is no other subdirectory beginning with `prog`. Tab completion works with any command you type into the IPython terminal. Try it out! It will make your life more wonderful.

A related shortcut involves the ↑ key. If you type a command, say `cd` and then press the ↑ key, IPython will complete the `cd` command with the last instance of that command. Thus, when you launch

Launching Python

IPython, you can use this shortcut to take you to the directory you used when you last ran IPython.

You can also simply press the ↑ key, which will recall the most recent command. Repeated application of the ↑ key scrolls through the most recent commands in reverse order. The ↓ key can be used to scroll in the other direction.

2.4.4 Recap of commands

Let's recap the most useful commands introduced above:

`%pwd`: (**p**rint **w**orking **d**irectory) Prints the path of the current directory.

`%ls`: (**l**ist) Lists the names of the files and directories located in the current directory.

`%mkdir` *filename*: (**m**a**k**e **dir**ectory) Makes a new directory *filename*.

`%cd` *directoryname*: (**c**hange **d**irectory) Changes the current directory to *directoryname*. Note: for this to work, *directoryname* must be a subdirectory in the current directory. Typing `%cd` changes to the home directory of your computer. Typing `%cd ..` moves the console one directory up in the directory tree.

`%clear`: Clears the IPython screen of previous commands.

`%run` *filename*: Runs (executes) a Python script. Described later in §2.7.1.

Tab completion: Provides convenient shortcuts, with or without the arrow keys, for executing commands in the IPython shell.

2.5 Interactive Python as a Calculator

You can use the IPython shell to perform simple arithmetic calculations. For example, to find the product 3×15, you type `3*15` at the `In` prompt and press RETURN:

```
In [1]: 3*15
Out[1]: 45
```

Python returns the correct product, as expected. You can do more complicated calculations:

```
In [2]: 6+21/3
Out[2]: 13.0
```

Let's try some more arithmetic:

```
In [3]: (6+21)/3
Out[3]: 9.0
```

Notice that the effect of the parentheses in `In [3]: (6+21)/3` is to cause the addition to be performed first and then the division. Without the parentheses, Python will always perform the multiplication and division operations *before* performing the addition and subtraction operations. The order in which arithmetic operations are performed is the same as for most calculators: exponentiation first, then multiplication or division, then addition or subtraction, then left to right.

2.5.1 Binary arithmetic operations in Python

Table 2.1 below lists the binary arithmetic operations in Python. It has all the standard binary operators for arithmetic, plus a few you may not have seen before.

Operation	Symbol	Example	Output
addition	+	12+7	19
subtraction	-	12-7	5
multiplication	*	12*7	84
division	/	12/7	1.714285
floor division	//	12//7	1
remainder	%	12%7	5
exponentiation	**	12**7	35831808

Table 2.1 Binary operators.

"Floor division," designated by //, means divide and keep only the integer part without rounding. "Remainder," designated by the symbol %, gives the remainder after floor division.

2.5.2 Types of numbers

There are three different types of numbers in Python: integers, floating point numbers, and complex numbers.

Integers in Python are simply, as their name implies, integers.

Launching Python

They can be positive or negative and can be arbitrarily long. In Python, a number is automatically treated as an integer if it is written without a decimal. This means that 23, written without a decimal point, is an integer and 23., written with a decimal point, is a floating point number. Here are some examples of integer arithmetic:

```
In [4]: 12*3
Out[4]: 36

In [5]: 4+5*6-(21*8)
Out[5]: -134

In [6]: 11/5
Out[6]: 2.2

In [7]: 11//5     # floor divide
Out[7]: 2

In [8]: 9734828*79372
Out[8]: 772672768016
```

For the binary operators +, -, *, and //, the output is an integer if the inputs are integers. The output of the division operator / is a floating point as of version 3 of Python. If an integer output is desired when two integers are divided, the floor division operator // must be used.

Floating point numbers are essentially rational numbers and can have a fractional part; integers, by their very nature, have no fractional part. In most versions of Python running on PCs or Macs, floating point numbers go between approximately $\pm 2 \times 10^{-308}$ and $\pm 2 \times 10^{308}$. Here are some examples of floating point arithmetic:

```
In [9]: 12.*3
Out[9]: 36.0

In [10]: 5**0.5
Out[10]: 2.23606797749979

In [11]: 11./5.
Out[11]: 2.2

In [12]: 11.//5.
Out[12]: 2.0

In [13]: 11.%5.
Out[13]: 1.0
```

```
In [14]:  6.022e23*300.
Out[14]:  1.8066e+26
```

Note that the result of any operation involving only floating point numbers as inputs is another floating point number, even in the cases where the floor division `//` or remainder `%` operators are used. The last output also illustrates an alternative way of writing floating point numbers as a mantissa followed by `e` or `E` followed by a power of 10: so `1.23e-12` is equivalent to 1.23×10^{-12}.

We also used the exponentiation operator `**` to find the square root of 5 by using a fractional power of `0.5`.

Complex numbers are written in Python as a sum of a real and imaginary part. For example, the complex number $3 - 2i$ is represented as `3-2j` in Python where `j` represents $\sqrt{-1}$. Here are some examples of complex arithmetic:

```
In [15]:  (2+3j)*(-4+9j)
Out[15]:  (-35+6j)

In [16]:  (2+3j)/(-4+9j)
Out[16]:  (0.1958762886597938-0.3092783505154639j)

In [17]:  2.5-3j**2
Out[17]:  (11.5+0j)

In [18]:  (2.5-3j)**2
Out[18]:  (-2.75-15j)
```

Notice that you need to enclose the real and imaginary parts of a complex number in parentheses if you want operators to operate on the entire complex number.

If you multiply an integer by a floating point number, the result is a floating point number. If you multiply a floating point number by a complex number, the result is a complex number. Python promotes the result to the most complex of the inputs.

2.5.3 Important note on integer division in Python

One peculiarity of all versions of Python prior to version 3 is that dividing two integers by each other yields the "floor division" result—another integer. Therefore `3/2` yields `1` whereas `3./2` or `3/2.` or `3./2.` all yield `1.5`. Starting with version 3 of Python, all of the above expressions, including `3/2` yield `1.5`. Fortunately, we are using version 3 of Python so `3/2` yields `1.5`. However, you may

Launching Python

run into other installations of Python that use version 2, so you should be aware of this point. You can force versions of Python prior to version 3 to divide integers like version 3 does by typing `from __future__ import division` at the beginning of an IPython session. You only need to type it once and it works for the entire session.

2.6 Variables

2.6.1 Names and the assignment operator

A variable is a name that is used to store data. It can be used to store different kinds of data, but here we consider a simple case where the data is a single numerical value. Here are a few examples:

`In [1]: a = 23`

`In [2]: p, q = 83.4, 2**0.5`

The equal sign "=" is the assignment operator. In the first statement, it creates an integer `a` and assigns it a value of 23. In the second statement it creates two floating point numbers `p` and `q` and assigns them the values of 83.4 and 1.4142135623730951, respectively.

To be more precise, when we write `a = 5`, Python creates an integer *object* and assigns it a value of 5. When we write `p, q = 83.4, 2**0.5`, Python creates two floating point objects and assigns each its own value. Note that Python stores the *numerical value*, not the expression used to generate it. Thus, `q` is assigned the 17-digit number 1.4142135623730951 generated by evaluating the expression `2**0.5`, *not* with $\sqrt{2}$. (Actually the value of `q` is stored as a binary, base 2, number using scientific notation with a mantissa and an exponent.)

In the previous paragraph, we said that the assignment operator creates an *object*. We will have much more to say about Python objects later on, with an explicit discussion of what they are in §3.5. But for now, it suffices to say that variable objects, like `a`, `p`, and `q` defined above, contain both data values, such as 23, 83.4, and 1.4142135623730951, as well as information about the data, such as the data type. For these cases, that data stored in the variable `a` is a single integer while the data stored in `p` is a floating point number, as is `q`.

The assignment variable works from right to left; that is, it assigns the value of the number on the right to the variable name on the left.

Therefore, the statement "5=a" makes no sense in Python. The assignment operator "=" in Python is not equivalent to the equals sign "=" we are accustomed to in algebra.

The assignment operator can be used to increment or change the value of a variable.

```
In [3]: a = a+1

In [4]: a
Out[4]: 24
```

The statement, a = a+1 makes no sense in algebra, but in Python (and most computer languages), it makes perfect sense: it means "add 1 to the current value of a and assign the result to a." This construction appears so often in computer programming that there is a special set of operators to perform such changes to a variable: +=, -=, *=, and /=. Here are some examples of how they work:

```
In [5]: c , d = 4, 7.92

In [6]: c += 2

In [7]: c
Out[7]: 6

In [8]: c *= 3

In [9]: c
Out[9]: 18

In [10]: d /= -2

In [11]: d
Out[11]: -3.96

In [12]: d -= 4

In [13]: d
Out[13]: -7.96
```

By the way, %=, **=, and //=, are also valid operators. Verify that you understand how the above operations work.

2.6.2 Legal and recommended variable names

Variable names in Python must start with a letter or an underscore "_" and can be followed by as many alphanumeric characters as you like, including the underscore character "_". Spaces are not allowed in variable names. No other character that is not a letter or a number is permitted.

Although variable names can start with the underscore character, you should avoid doing so in most cases. Variables beginning with an underscore are generally reserved for special cases, with which we need not concern ourselves here.

Recall that Python is *case sensitive*, so the variable b is distinct from the variable B.

We recommend giving your variables descriptive names as in the following calculation:

```
In [14]: distance = 34.

In [15]: time_traveled = 0.59

In [16]: velocity = distance/time_traveled

In [17]: velocity
Out[17]: 57.6271186440678
```

The variable names distance, time_traveled, and velocity immediately remind you of what is being calculated here. This is good practice. But so is keeping variable names reasonably short, so don't go nuts!

2.6.3 Reserved words in Python

There are also some names or words that are reserved by Python for special purposes. You must avoid using these names as variables, which are provided in Table 2.2 for your reference.

False	class	finally	is	return
None	continue	for	lambda	try
True	def	from	nonlocal	while
and	del	global	not	with
as	elif	if	or	yield
assert	else	import	pass	
break	except	in	raise	

Table 2.2   Reserved names in Python.

2.7 Script Files and Programs

Performing calculations in the IPython shell is handy if the calculations are short. But calculations quickly become tedious when they are more than a few lines long. If you discover you made a mistake at some early step, for example, you may have to go back and retype all the steps subsequent to the error. Having code saved in a file means you can just correct the error and rerun the code without having to retype it. Saving code can also be useful if you want to reuse it later, perhaps with different inputs.

For these and many other reasons, we save code in computer files. We call the sequence of commands stored in a file a *script* or a *program* or sometimes a *routine*. Programs can become quite sophisticated and complex. Here we are only going to introduce the simplest features of programming by writing a very simple script. Later, we will introduce some of the more advanced features of programming.

To write a script you need a text editor. In principle, any text editor will do, but it's more convenient to use an editor that was designed for the task. We are going to use the **Editor** of the Spyder IDE (see Fig. 2.1). The Spyder Editor, like most good programming editors, provides syntax highlighting, which color codes keywords, comments, and other features of the Python syntax according to their function, and thus makes it easier to read the code and easier to spot programming mistakes. The Spyder Editor also provides syntax checking, much like a spell-checker in a word processing program, that identifies many coding errors. This can greatly speed the coding process. Tab completion also works in the Editors.

2.7.1 First scripting example: The Editor pane

Let's work through an example to see how scripting works. Suppose you are going on a road trip and you would like to estimate how long the drive will take, how much gas you will need, and the cost of the gas. It's a simple calculation. As inputs, you will need the distance of the trip, your average speed, the cost of gasoline, and the mileage (average miles per gallon) of your car.

Writing a script to do these calculations is straightforward. First, launch Spyder. You should see a tab with the word `untitled` at the top left of the Editor Pane (see Fig. 2.1). If you don't, go to the `File` menu and select `New File`. Use the mouse to place your cursor at the top of

Launching Python

the Editor pane. Enter the following code and *save the code* in a file called `myTrip.py`. Place the file in the directory `programs` that you created earlier (see §2.4.1). This stores your script (or program) on your computer's disk. The exact name of the file is not important but the extension `.py` is essential. It tells the computer, and more importantly Python, that this is a Python program.

Code: chapter2/programs/myTrip.py

```
1  # Calculates time, gallons of gas used, and cost of
2  # gasoline for a trip
3  distance = 400.          # miles
4  mpg = 30.                # car mileage
5  speed = 60.              # average speed
6  costPerGallon = 2.85     # price of gas
7
8  time = distance/speed
9  gallons = distance/mpg
10 cost = gallons*costPerGallon
```

The number (or hash) symbol # is the "comment" character in Python; anything on a line following # is ignored when the code is executed. Judicious use of comments in your code will make your code much easier to understand days, weeks, or months after the time you wrote it. Use comments generously. For aesthetic reasons, the comments on different lines have been aligned. This isn't necessary. The trailing spaces needed to align the comments have no effect on the running of the code.

Now you are ready to run the code. Before doing so, you first need to use the IPython console to move to the `programs` directory where the file containing the code resides. That is, from the IPython console, use the `cd` command to move to the `programs` directory. For example, you might type

```
In [1]: %cd ~/Documents/programs/
```

To *run* or *execute* a script, simply type `%run` *filename*, which in this case means type `%run myTrip.py` (if you have IPython's Automagic switch turned on, as described in §2.4.1, you can omit the percent sign and just type `run` *filename*). When you run a script, Python simply executes the sequence of commands in the order they appear.

```
In [2]: %run myTrip.py
```

Once you have run the script, you can see the values of the variables calculated in the script simply by typing the name of the variable. IPython responds with the value of that variable.

```
In [3]: time
Out[3]: 6.666666666666667

In [4]: gallons
Out[4]: 13.333333333333334

In [5]: cost
Out[5]: 38.0
```

You can change the number of digits IPython displays using the magic command %precision:

```
In [6]: %precision 2
Out[6]: ' %.2f'

In [7]: time
Out[7]: 6.67

In [8]: gallons
Out[8]: 13.33

In [9]: cost
Out[9]: 38.00
```

Typing %precision returns IPython to its default state; %precision %e causes IPython to display numbers in exponential format (scientific notation).

Note about printing

If you want your script to return the value of a variable (that is, print the value of the variable to your computer screen), use the print function. For example, at the end of our script, if we include the code

```
print(time)
print(gallons)
print(cost)
```

the script will return the values of the variables time, gallons, and cost that the script calculated. We will discuss the print function in much greater detail, as well as other methods for data output, in Chapter 4.

2.8 Python Modules

The Python computer language consists of a "core" language plus a vast collection of supplementary software that is contained in **mod-**

Launching Python

ules (or packages, which are collections of modules—we'll not fuss about the distinction here). Many of these modules come with the standard Python distribution and provide added functionality for performing computer system tasks. Other modules provide more specialized capabilities that not every user may want. You can think of these modules as a kind of library from which you can borrow according to your needs. You gain access to a module using the `import` command, which we introduce in the next section.

We will need four Python modules that are not part of the core Python distribution, but are nevertheless widely used for scientific computing. The four modules are:

NumPy is the standard Python package for scientific computing with Python. It provides the all-important NumPy `array` data structure, which is at the very heart of NumPy. It also provides tools for creating and manipulating arrays, including indexing and sorting, as well as basic logical operations and element-by-element arithmetic operations like addition, subtraction, multiplication, division, and exponentiation. It includes the basic mathematical functions of trigonometry, exponentials, and logarithms, as well as a vast collection of special functions (Bessel functions, *etc.*), statistical functions, and random number generators. It also includes a large number of linear algebra routines that overlap with those in SciPy, although the SciPy routines tend to be more complete. You can find more information about NumPy at http://docs.scipy.org/doc/numpy/reference/index.html.

SciPy provides a wide spectrum of mathematical functions and numerical routines for Python. SciPy makes extensive use of NumPy arrays so when you import SciPy, you should always import NumPy too. In addition to providing basic mathematical functions, SciPy provides Python "wrappers" for numerical software written in other languages, like Fortran, C, or C++. A "wrapper" provides a transparent easy-to-use Python interface to standard numerical software, such as routines for doing curve fitting and numerically solving differential equations. SciPy greatly extends the power of Python and saves you the trouble of writing software in Python that someone else has already written and optimized in some other language. You can find more information about SciPy at http://docs.scipy.org/doc/scipy/reference/.

matplotlib is the standard Python package for making two- and three-dimensional plots. matplotlib makes extensive use of NumPy arrays. You will make all of your plots in Python using this package. You can find more information about matplotlib at http://matplotlib.sourceforge.net/.

Pandas is a Python package providing a powerful set of data analysis tools. It uses data structures similar to those used in a spreadsheet program like Excel, and allows you to manipulate data in ways similar to what is done using spreadsheets. You can find more information about Pandas at http://pandas.pydata.org/.

We will use these four modules extensively and therefore will provide introductions to their capabilities as we develop Python. The links above provide much more extensive information and you will certainly want to refer to them from time to time.

2.8.1 Python modules and functions: A first look

Because the modules listed above, NumPy, SciPy, matplotlib, and Pandas, are not part of core Python, they need to be imported before we can gain access to their functions and data structures. Here, we show how to import the NumPy module and use some of its functions. We defer introducing NumPy arrays, mentioned in the previous section, until §3.3.

We gain access to the NumPy package using Python's `import` statement:

```
In [1]: import numpy
```

After running this statement, we have access to all the functions and data structures of NumPy. For example, we can now access NumPy's sine function as follows:

```
In [2]: numpy.sin(0.5)
Out[2]: 0.479425538604203
```

In this simple example, the `sin` function has one argument, here `0.5`, and the function returns the sine of that argument, which is assumed to be expressed in units of radians.

Note that we had to put the prefix `numpy` dot before the name of the actual function name `sin`. This tells Python that the `sin` function is part of the NumPy module that we just imported.

There is another Python module called `math` that also has a sine function. We can import the math module just like we imported the NumPy module:

```
In [3]: import math
In [4]: math.sin(0.5)
Out[4]: 0.479425538604203
```

These two sine functions are not the same function, even though in this case they give the same answer. Consider, for example, what happens if we ask each function to find the sine of a complex number:

```
In [5]: numpy.sin(3+4j)
Out[5]: (3.853738037919377-27.016813258003932j)

In [6]: math.sin(3+4j)
---------------------------------------------------------------
TypeError                   Traceback (most recent call last)
<ipython-input-24-b48edfeaf02a> in <module>()
----> 1 math.sin(3+4j)

TypeError: can't convert complex to float
```

The NumPy sine function works just fine and returns a complex result. By contrast, the math sine function returns a error message because it does not accept a complex argument. In fact, the math sine function accepts only a single real number as an argument while the numpy sine function accepts real and complex NumPy arrays, which we introduce in §3.3, as arguments. For single real arguments, the math sine function executes faster than the numpy function, but the difference in execution speed is not noticeable in most cases.

The important lesson here is to appreciate how Python allows you to extend its capabilities by importing additional packages, while at the same time keeping track of where these capabilities come from using the prefix dot syntax. By using different prefixes, each module maintains its own *namespace*, that is, its own separate dictionary of names, so that functions with the same name in different packages do not clash.

If you are using a lot of NumPy functions, writing out numpy dot before each function can be a little verbose. Python allows you to define an abbreviation for the prefix when you import a library. Here we show how to do it for NumPy:

```
In [7]: import numpy as np

In [8]: np.sin(0.5)
Out[8]: 0.47942553860420301
```

The statement `import numpy as np` imports and assigns the abbreviation `np` for `numpy`. In principle, you can use any abbreviation you wish. However, it's common practice to use `np` for the NumPy module. You are strongly encouraged to abide by this practice so that others reading your code will recognize what you are doing.

2.8.2 Some NumPy functions

NumPy includes an extensive library of mathematical functions. In Table 2.3, we list some of the most useful ones. A complete list is available at https://docs.scipy.org/doc/numpy/reference/.

Function	Description
sqrt(x)	square root of x
exp(x)	exponential of x, i.e., e^x
log(x)	natural log of x, i.e., $\ln x$
log10(x)	base 10 log of x
degrees(x)	converts x from radians to degrees
radians(x)	converts x from degrees to radians
sin(x)	sine of x (x in radians)
cos(x)	cosine x (x in radians)
tan(x)	tangent x (x in radians)
arcsin(x)	Arc sine (in radians) of x
arccos(x)	arc cosine (in radians) of x
arctan(x)	arc tangent (in radians) of x
fabs(x)	absolute value of x
math.factorial(n)	$n!$ of an integer
round(x)	rounds a float to nearest integer
floor(x)	rounds a float *down* to nearest integer
ceil(x)	rounds a float *up* to nearest integer
sign(x)	-1 if $x < 0$, $+1$ if $x > 0$, 0 if $x = 0$

Table 2.3 Some NumPy math functions.

The argument of these functions can be a number or any kind of expression whose output produces a number. All of the following expressions are legal and produce the expected output:

```
In [9]: np.log(np.sin(0.5))
Out[9]: -0.73516668638531424

In [10]: np.log(np.sin(0.5)+1.0)
```

```
Out[10]: 0.39165386283471759

In [11]: np.log(5.5/1.2)
Out[11]: 1.5224265354444708
```

Here, we have demonstrated functions with one input and one output. In general, Python functions have multiple inputs and multiple outputs. We will discuss these and other features of functions later when we take up functions in the context of user-defined functions.

2.8.3 Scripting Example 2

Let's try another problem. Suppose you want to find the distance between two Cartesian coordinates (x_1, y_1, z_1) and (x_2, y_2, z_2). The distance is given by the formula

$$\Delta r = \sqrt{(x_2 - x_1)^2 + (y_2 - y_1)^2 + (z_2 - z_1)^2}$$

Now let's write a script to do this calculation and save it in a file called twoPointDistance.py.

Code: chapter2/programs/twoPointDistance.py

```
1  # Calculates the distance between two 3d Cartesian
2  # coordinates
3  import numpy as np
4
5  x1, y1, z1 =  23.7, -9.2, -7.8
6  x2, y2, z2 =  -3.5,  4.8,  8.1
7
8  dr = np.sqrt((x2-x1)**2 + (y2-y1)**2 + (z2-z1)**2)
```

We have introduced extra spaces into some of the expressions to improve readability. They are not necessary; where and whether you include them is largely a matter of taste.

Because we will need the square root function of NumPy, the script imports NumPy before doing anything else. If you leave out the "`import numpy as np`" line or remove the `np` dot in front of the `sqrt` function, you will get the following error message

```
Traceback (most recent call last):
    ...
  File ".../twoPointDistance.py", line 8, in <module>
    dr = sqrt((x2-x1)**2 + (y2-y1)**2 + (z2-z1)**2)

NameError: name "sqrt" is not defined
```

Now, with the `import numpy as np` statement, we can run the script.

```
In [10]: %run twoPointDistance.py
```

```
In [11]: dr
Out[11]: 34.48
```

The script works as expected.

2.8.4 Different ways of importing modules

There are different ways that you can import modules in Python.

Importing an entire module

Usually we import entire modules using the `import` statement or the `import ... as ...` statement that we introduced for the Math and NumPy libraries:

```
import math
import numpy as np
```

Importing part of a module

You can also import a single function or subset of functions from a module without importing the entire module. For example, suppose you wanted to import just the `log` function from NumPy. You could write

```
from numpy import log
```

To use the `log` function in a script, you would write

```
a = log(5)
```

which would assign the value `1.6094379124341003` to the variable `a`. If you wanted to import the three functions, `log`, `sin`, and `cos`, you would write

```
from numpy import log, sin, cos
```

Imported in this way, you would use them without any prefix as the functions are imported into the general namespace of the program. In general, we do not recommend using `from` *module* `import` *functions* way of importing functions. When reading code, it makes it harder to determine from which modules functions are imported, and can lead to clashes between similarly named functions from different modules. Nevertheless, we do use this form sometimes and, more importantly, you will see the form used in programs you encounter on the web and elsewhere so it is important to understand the syntax.

Blanket importing of a module

There is yet another way of importing an entire module by writing

```
from numpy import *
```

This imports the entire module, in this case NumPy, into the general namespace and allows you to use all the functions in the module without a prefix. If you import two different libraries this way in the same script, then it's impossible to tell which functions come from which library by just looking at the script. You also have the aforementioned problem of clashes between libraries, so you are strongly advised not to import this way in a script or program.

There is one possible exception to this advice, however. When working in the IPython shell, you generally just want to try out a function or very small snippet of code. You usually are not saving this code in a script; it's disposable code, never to be seen or used again. In this case, it can be convenient to not have to write out the prefixes. If you like to operate this way, then type pylab at the IPython prompt. This imports NumPy and matplotlib as follows:

```
from numpy import *
from matplotlib.pyplot import *
```

In Appendix A, we suggest that you set up Spyder so that it does not automatically launch the IPython shell in "pylab" mode. While you are learning Python, it's important that you learn which functions belong to which modules. After you become more expert in Python, you can decide if you want to work in an IPython shell in pylab mode.[1]

In this text, we **do not** operate our IPython shell in "pylab" mode. That way, it is always clear to you where the functions we use come from.

Whether you choose to operate your IPython shell in pylab mode or not, the NumPy and matplotlib libraries (as well as other libraries) are not available in the scripts and programs you write in the Editor Pane unless you explicitly import these modules, which you would do by writing

---

[1] Some programmers consider such advice sacrilege. Personally, I sometimes find pylab mode to be convenient for my workflow. You can decide if it suits you.

```
import numpy as np
import matplotlib.pyplot as plt
```

2.9 Getting Help: Documentation in IPython

Help is never far away when you are running the IPython shell. To obtain information on any valid Python or NumPy function, and many matplotlib functions, simply type help(*function*), as illustrated here

```
In [1]: help(range)

Help on class range in module builtins:

class range(object)
class range(object)
 |  range(stop) -> range object
 |  range(start, stop[, step]) -> range object
 |
 |  Return an object that produces a sequence of integers
 |  from start (inclusive) to stop (exclusive) by step.
 |  range(i, j) produces i, i+1, i+2, ..., j-1. start defaults
 |  to 0, and stop is omitted! range(4) produces 0, 1, 2, 3.
 |  These are exactly the valid indices for a list of 4
 |  elements. When step is given, it specifies the increment
 |  (or decrement).
```

Often, the information provided can be quite extensive and you might find it useful to clear the IPython window with the %clear command so you can easily scroll back to find the beginning of the documentation. You may have also noticed that when you type the name of a function plus the opening parenthesis, IPython displays a small window describing the basic operation of that function.

One nice feature of the *Spyder* IDE is that if you place the cursor next to or within a function name and press 'cmd-I' (Mac) or 'ctrl-I' (PC), the web documentation for the function is displayed in the Help Pane.

2.10 Stand-alone IPython

You don't need the Spyder IDE to run IPython. You can run the IPython shell on its own. To do so, you first need to launch a terminal application. If you are running the macOS, launch the Terminal

Launching Python

application, which you can find in the Application/Utilities folder. If you are running Windows, launch the Anaconda Prompt app under the Anaconda (or Anaconda3) menu in the Start menu.

Once you launch a terminal application, you should type `ipython` at the prompt. This will bring up an IPython terminal window with the usual IPython prompt.[2]

From there, you can execute any valid Python or IPython command or function.

Note, however, that the NumPy module is not automatically loaded. Therefore, you will need to type `import numpy as np` and then use the `np` dot prefix to access any NumPy functions you may wish to use.

You can run a Python script by typing `%run` followed by the script file name. For example, to run the the script `myTrip.py`, you would type:

```
In [1]: %run myTrip.py
```

Of course, for this to work, the script `myTrip.py` must be in the working director of IPython. If this is not the case, you can use the `cd` command to navigate to the directory in which the script `myTrip.py` is found.

You can close the IPython shell by typing `quit()`.

2.10.1 Writing Python scripts in a text editor

You can write Python scripts using any plain text editor. No special editor or IDE is required. Some editors, however, automatically recognize any file whose names ends in the suffix `.py` as a Python file. For example, the text editor programs Notepad++ (for PCs) or BBEdit (for Macs) recognize Python files. These editors are nice because the color code the Python syntax and provide other useful features. There are many others that work too.

One particularly interesting text editor is called Atom. It's free and available for for Macs, PCs, and Linux machines. It's highly configurable and can be set up to work as a very effective Python editor.

---

[2] You can launch the standard standard Python interactive shell, which is distinct from the IPython shell, simply by typing `python` and pressing RETURN. You will get the standard Python prompt >>>. Of course, you won't get the functionality of the IPython shell with its magic commands, as described in §2.4.1. To close the Python shell, type `quit()`.

2.11 Programming Errors

Now that you have a little experience with Python and computer programming, it's time for an important reminder: **Programming is a detail-oriented activity.** To be good at computer programming, to avoid frustration when programming, you must pay attention to details. A misplaced or forgotten comma or colon can keep your code from working. Note that I did not say it can "keep your code from working *well*"; it can keep your code from working at all! Worse still, little errors can make your code give erroneous answers, where your code appears to work, but in fact does not do what you intended it to do! So pay attention to the details!

2.11.1 Pyflakes

One way to avoid making errors is to use a syntax checker. Fortunately, a syntax checker is built into the Spyder IDE This is an *enormous asset* when writing code and one of the best reasons for using an IDE.

A syntax-checking program called *Pyflakes* runs in the background when you are editing a Python program using Spyder. If there is an error in your code, Pyflakes flags the error.

In Spyder, a red circle ● appears to the left of the line where Pyflakes thinks the error occurs. Sometimes, the error actually occurs in the previous line, so look around. A yellow triangle ▲ appears to the left of the line where Pyflakes thinks that the coding style doesn't conform to the PEP 8 standard;[3] it's not an error, just a coding style violation, which you can heed or ignore. Passing your mouse pointer over the red or yellow icon brings up a *Code Analysis* box with a brief message describing the error or style violation.

In this text, we have mostly heeded the PEP 8 style guidelines (see https://www.python.org/dev/peps/pep-0008). We advise you to do the same. PEP 8 is Python's style guide and its aim is to enhance code readability. The simplest way to learn the style guidelines is to heed the messages associated with the yellow triangle icons. Doing so will make the yellow icons disappear and will generally render your code more consistent and readable.

---

[3] PEP stands for *Python Enhancement Proposal*.

2.11.2 Error checking

This raises a second point: sometimes your code will run but give the wrong answer because of a programming error or because of a more subtle error in your algorithm, even though there may be nothing wrong with your Python syntax. The program runs; it just gives the wrong answer. For this reason, it is important to test your code to make sure it is behaving properly. Test it to make sure it gives the correct answers for cases where you already know the correct answer or where you have some independent means of checking it. Test it in limiting cases, that is, for cases that are at the extremes of the sets of parameters you will employ. *Always test your code; this is a cardinal rule of programming.*

2.12 Exercises

1. A ball is thrown vertically up in the air from a height h_0 above the ground at an initial velocity v_0. Its subsequent height h and velocity v are given by the equations

$$h = h_0 + v_0 t - \frac{1}{2}gt^2$$
$$v = v_0 - gt$$

where $g = 9.8$ is the acceleration due to gravity in m/s². Write a script that finds the height h and velocity v at a time t after the ball is thrown. Start the script by setting $h_0 = 1.6$ (meters) and $v_0 = 14.2$ (m/s) and have your script print out the values of height and velocity. Then use the script to find the height and velocity after 0.5 seconds. Then modify your script to find them after 2.0 seconds.

2. Write a script that defines the variables $V_0 = 10, a = 2.5$, and $z = 4\frac{1}{3}$, and then evaluates the expression

$$V = V_0 \left(1 - \frac{z}{\sqrt{a^2 + z^2}}\right).$$

Then find V for $z = 8\frac{2}{3}$ and print it out (see *Note about printing* on page 18). Then find V for $z = 13$ by changing the value of z in your script.

3. Write a single Python script that calculates the following expressions:

   (a) $a = \dfrac{2 + e^{2.8}}{\sqrt{13} - 2}$

   (b) $b = \dfrac{1 - (1 + \ln 2)^{-3.5}}{1 + \sqrt{5}}$

   (c) $c = \sin\left(\dfrac{2 - \sqrt{2}}{2 + \sqrt{2}}\right)$

   After running your script in the IPython shell, typing `a`, `b`, or `c` at the IPython prompt should yield the value of the expressions in (a), (b), or (c), respectively.

4. A quadratic equation with the general form

   $$ax^2 + bx + c = 0$$

   has two solutions given by the quadratic formula

   $$x = \dfrac{-b \pm \sqrt{b^2 - 4ac}}{2a}.$$

   (a) Given a, b, and c as inputs, write a script that gives the numerical values of the two solutions. Write the constants a, b, and c as floats, and show that your script gives the correct solutions for a few test cases when the solutions are real numbers, that is, when the discriminant $b^2 - 4ac \geq 0$. Use the `print` function in your script, discussed at the end of §2.7.1, to print out your two solutions.

   (b) Written this way, however, your script gives an error message when the solutions are complex. For example, see what happens when $a = 1$, $b = 2$, and $c = 3$. You can fix this using statements in your script like `a = a+0j` after setting a to some float value. Thus, you can make the script work for any set of real inputs for a, b, and c. Again, use the `print` function to print out your two solutions.

5. Write a program to calculate the perimeter p of an n-gon inscribed inside a sphere of diameter 1. Find p for $n = 3, 4, 5, 100, 10{,}000$, and $1{,}000{,}000$. Your answers should be

n	p	n	p
3	2.59807621135	100	3.14107590781
4	2.82842712475	10,000	3.14159260191
5	2.93892626146	1,000,000	3.14159265358

CHAPTER 3

Strings, Lists, Arrays, and Dictionaries

> *In this chapter you learn about* **data structures**, *which Python uses to store and organize numerical, alphabetical, and other types of information. The variables introduced in the previous chapter are a very simple kind of data structure. Here we introduce several more data structures that prove useful in programming, including* **strings, lists, tuples,** *and* **dictionaries,** *which are all part of core Python. We also introduce* **NumPy arrays,** *which are very useful for storing and manipulating scientific data. We introduce a powerful technique called* **slicing,** *which allows you to extract and manipulate sections of data contained in lists, tuples, and NumPy arrays. Finally, we introduce some basic ideas about* **objects,** *which are central to the underlying structure and functioning of Python.*

The most important data structure for scientific computing in Python is the **NumPy array**. NumPy arrays are used to store lists of numerical data and to represent vectors, matrices, and even tensors. NumPy arrays are designed to handle large data sets efficiently and with a minimum of fuss. The NumPy library has a large set of routines for creating, manipulating, and transforming NumPy arrays. NumPy functions, like `sqrt` and `sin`, are designed specifically to work with NumPy arrays. Core Python has an array data structure, but it's not nearly as versatile, efficient, or useful as the NumPy array. We will not be using Python arrays at all. Therefore, whenever we refer to an "array," we mean a "NumPy array." We discuss NumPy arrays in §3.3.

Lists are another data structure, similar to NumPy arrays, but unlike NumPy arrays, lists are a part of core Python. Lists have a variety of uses. They are useful, for example, in various bookkeeping tasks that arise in computer programming. Like arrays, they are sometimes used to store data. However, lists do not have the specialized properties and tools that make arrays so powerful for scientific computing. Therefore, we usually prefer arrays to lists for working with scientific data, but there are some circumstances for which using lists is preferable, even for scientific computing. And for other tasks, lists work just fine. We will use them frequently. We discuss them in §3.2.

Strings are lists of keyboard characters as well as other characters not on your keyboard. They are not particularly interesting in scientific computing, but they are nevertheless necessary and useful. Texts on programming with Python typically devote a good deal of time and space to learning about strings and how to manipulate them. Our uses of them are rather modest, however, so we take a minimalist's approach and only introduce a few of their features. We discuss strings in §3.1.

Dictionaries are like lists, but the elements of dictionaries are accessed in a different way than for lists. The elements of lists and arrays are numbered consecutively, and to access an element of a list or an array, you simply refer to the number corresponding to its position in the sequence. The elements of dictionaries are accessed by "keys," which can be program strings or (arbitrary) integers (in no particular order). Dictionaries are an important part of core Python. We introduce them in §3.4.

3.1 Strings

Strings are lists of characters. Any character that you can type from a computer keyboard, plus a variety of other characters, can be elements in a string. Strings are created by enclosing a sequence of characters within a pair of single or double quotes. Examples of strings include `"Marylyn"`, `'omg'`, `"good_bad_#5f>"`, `"{0:0.8g}"`, and `'We hold these truths ...'`. Caution: the defining quotes must *both* be single or *both* be double quotes when defining a given string. But you can use single quotes to define one string and double quotes to define the next string; it's up to you and has no consequence.

Strings can be assigned variable names

```
In [1]: a = "My dog's name is"
In [2]: b = "Bingo"
```

Note that we used double quotes to define the string a, so that we could use the apostrophe (single quote) in `dog's`. Strings can be concatenated using the "+" operator:

```
In [3]: c = a + " " + b
In [4]: c
Out[4]: "My dog's name is Bingo"
```

Strings, Lists, Arrays, and Dictionaries

In forming the string c, we concatenated *three* strings, a, b, and a *string literal*, in this case a space " ", which is needed to provide a space to separate string a from b.

You will use strings for different purposes: labeling data in data files, labeling axes in plots, formatting numerical output, requesting input for your programs, as arguments in functions, *etc.*

Because numbers—digits—are also alpha numeric characters, strings can be made up of numbers:

```
In [5]: d = "927"
In [6]: e = 927
```

The variable d is a string while the variable e is an integer. If you try to add them by writing d+e, you get an error. However, if you type d + str(e) or int(d) + e, you get sensible, but different, results. Try them out!

3.2 Lists

Python has two data structures, *lists* and *tuples*, that consist of a list of one or more elements. The elements of lists or tuples can be numbers or strings, or both. Lists (we discuss tuples later in §3.2.3) are defined by a pair of *square* brackets on either end with individual elements separated by commas. Here are two examples of lists:

```
In [1]: a = [0, 1, 1, 2, 3, 5, 8, 13]
In [2]: b = [5., "girl", 2+0j, "horse", 21]
```

We can access individual elements of a list using the variable name for the list with an integer in square brackets:

```
In [3]: b[0]
Out[3]: 5.0

In [4]: b[1]
Out[4]: 'girl'

In [5]: b[2]
Out[5]: (2+0j)
```

The first element of b is b[0], the second is b[1], the third is b[2], and so on. Some computer languages index lists starting with 0, like Python and C, while others index lists (or things more-or-less equivalent) starting with 1 (like Fortran and MATLAB®). It's important to keep in mind that Python uses the former convention: lists are *zero-indexed*.

The last element of this array is `b[4]`, because b has 5 elements. The last element can also be accessed as `b[-1]`, no matter how many elements b has, and the next-to-last element of the list is `b[-2]`, *etc.* Try it out:

```
In [6]: b[4]
Out[6]: 21

In [7]: b[-1]
Out[7]: 21

In [8]: b[-2]
Out[8]: 'horse'
```

Individual elements of lists can be changed. For example:

```
In [9]: b
Out[9]: [5.0, 'girl', (2+0j), 'horse', 21]

In [10]: b[0] = b[0]+2

In [11]: b[3] = 3.14159

In [12]: b
Out[12]: [7.0, 'girl', (2+0j), 3.14159, 21]
```

Here we see that 2 was added to the previous value of `b[0]` and the string `'horse'` was replaced by the floating point number `3.14159`. We can also manipulate individual elements that are strings:

```
In [13]: b[1] = b[1] + "s & boys"

In [14]: b
Out[14]: [10.0, 'girls & boys', (2+0j), 3.14159, 21]
```

You can also add lists, but the result might surprise you:

```
In [15]: a
Out[15]: [0, 1, 1, 2, 3, 5, 8, 13]

In [16]: a+a
Out[16]: [0, 1, 1, 2, 3, 5, 8, 13, 0, 1, 1, 2, 3, 5,
    8, 13]

In [17]: a+b
Out[17]: [0, 1, 1, 2, 3, 5, 8, 13, 10.0, 'girls &
    boys', (2+0j), 3.14159, 21]
```

Strings, Lists, Arrays, and Dictionaries

Adding lists concatenates them, just as the "+" operator concatenates strings.

3.2.1 Slicing lists

You can access pieces of lists using the *slicing* feature of Python:

```
In [18]: b
Out[18]: [10.0, 'girls & boys', (2+0j), 3.14159, 21]

In [19]: b[1:4]
Out[19]: ['girls & boys', (2+0j), 3.14159]

In [20]: b[3:5]
Out[20]: [3.14159, 21]
```

You access a subset of a list by specifying two indices separated by a colon ":". This is a powerful feature of lists that we will use often. Here are a few other useful slicing shortcuts:

```
In [21]: b[2:]
Out[21]: [(2+0j), 3.14159, 21]

In [22]: b[:3]
Out[22]: [10.0, 'girls & boys', (2+0j)]

In [23]: b[:]
Out[23]: [10.0, 'girls & boys', (2+0j), 3.14159, 21]
```

Thus, if the left slice index is 0, you can leave it out; similarly, if the right slice index is the length of the list, you can leave it out also.

What does the following slice of an array give you?

```
In [24]: b[1:-1]
```

You can get the length of a list using Python's `len` function:

```
In [25]: len(b)
Out[25]: 5
```

You can also extract every second, third, or n^{th} element of a list. Here we extract every second and third element of a list starting at different points:

```
In [26]: b
Out[26]: [10.0, 'girls & boys', (2+0j), 3.14159, 21]

In [27]: b[0::2]
Out[27]: [10.0, (2+0j), 21]
```

```
In [28]: b[1::2]
Out[28]: ['girls & boys', 3.14159]

In [29]: b[0::3]
Out[29]: [10.0, 3.14159]

In [30]: b[1::3]
Out[30]: ['girls & boys', 21]

In [31]: b[2::3]
Out[31]: [(2+0j)]
```

3.2.2 The `range` function: Sequences of numbers

Because it turns out to be so useful, Python has a special function, `range`, that can be used to create a uniformly spaced sequence of integers. Its general form is `range([start,] stop[, step])`, where the arguments are all integers; those in square brackets are optional. In its simplest implementation it has only one argument, the `stop` argument:

```
In [32]: list(range(10))      # makes a list of 10
                              # integers from 0 to 9
Out[32]: [0, 1, 2, 3, 4, 5, 6, 7, 8, 9]

In [33]: list(range(3, 10))   # makes a list of
                              # integers from 3 to 9
Out[33]: [3, 4, 5, 6, 7, 8, 9]

In [34]: list(range(0, 10, 2))  # makes a list of 10
      # integers from 0 to 9 with an increment of 2
Out[34]: [0, 2, 4, 6, 8]
```

When two or three arguments are used, the first argument gives the first number of the list while the second argument ends the list, but is *not included in the list*. If the third argument is not included, it's taken to be 1.

```
In [35]:   a = list(range(4, 12))

In [36]: a
Out[36]: [4, 5, 6, 7, 8, 9, 10, 11]
```

You can use negative numbers with the `range` function, and increment the sequence with a number other than 1 using a third entry:

```
In [37]:   a = list(range(-5, 5, 2))
```

Strings, Lists, Arrays, and Dictionaries

```
In [38]: a
Out[38]: [-5, -3, -1, 1, 3]

In [39]: a = list(range(5, 0, -1))

In [40]: a
Out[40]: [5, 4, 3, 2, 1]
```

The range function makes an iterable sequence

In the examples above, we use the list function in conjunction with the range function because by itself the range function does not make a list. The range function simply generates a sequence of numbers one at a time. The only information stored in the range function is the current value of the sequence, the ending value, and the increment. By contrast, a list stores all the numbers in the list. For example, if we type range(10), Python does not return a list.

```
In [41]: range(10)
In [42]: range(0, 10)
```

Instead, the Python range(0, 10) function returns an *iterable sequence*, which saves a lot of memory when the sequence is very long. If this seems a bit vague or confusing at this point, don't fret. It will become clearer when we introduce for loops in §5.2.

3.2.3 Tuples

Next, a word about tuples: tuples are lists that are *immutable*. That is, once defined, the individual elements of a tuple cannot be changed. Whereas a list is written as a sequence of numbers enclosed in *square* brackets, a tuple is written as a sequence of numbers enclosed in *round* parentheses. Individual elements of a tuple are addressed in the same way as individual elements of lists are addressed, but those individual elements cannot be changed. All of this is illustrated by this simple example:

```
In [43]: c = (1, 1, 2, 3, 5, 8, 13)
In [44]: c[4]
Out[44]: 5

In [45]: c[4] = 7
Traceback (most recent call last):

  File "<ipython-input-2-7cb42185162c>", line 1,
```

```
in <module>
    c[4] = 7
```

```
TypeError: 'tuple' object does not support item
assignment
```

When we tried to change `c[4]`, the system returned an error because we are prohibited from changing an element of a tuple. Tuples offer some degree of safety when we want to define lists of immutable constants.

3.2.4 Multidimensional lists and tuples

We can also make multidimensional lists, or lists of lists. Consider, for example, a list of three elements, where each element in the list is itself a list:

```
In [40]: a = [[3, 9], [8, 5], [11, 1]]
```

Here we have a three-element list where each element consists of a two-element list. Such constructs can be useful in making tables and other structures. They also become relevant later on in our discussion of NumPy arrays and matrices, which we introduce in §3.3.

We can access the various elements of a list with a straightforward extension of the indexing scheme we have been using. The first element of the list a above is `a[0]`, which is `[3, 9]`; the second is `a[1]`, which is `[8, 5]`. The first element of `a[1]` is accessed as `a[1][0]`, which is 8, as illustrated below:

```
In [46]: a[0]
Out[46]: [3, 9]

In [47]: a[1]
Out[47]: [8, 5]

In [48]: a[1][0]
Out[48]: 8

In [49]: a[2][1]
Out[49]: 1
```

Multidimensional tuples work exactly like multidimensional lists, except they are immutable.

3.3 NumPy Arrays

The NumPy array is the real workhorse of data structures for scientific and engineering applications. The NumPy array, formally called ndarray in NumPy documentation, is similar to a list but where all the elements of the list are of the same type. The elements of a NumPy array, or simply an *array*, are usually numbers, but can also be Booleans, strings, or other objects. When the elements are numbers, they must all be of the same type. For example, they might be all integers or all floating point numbers.

3.3.1 Creating arrays (1-d)

NumPy has a number of functions for creating arrays. We focus on four (or five or six, depending on how you count!). The **first** of these, the array function, converts a list to an array:

```
In [1]: a = [0, 0, 1, 4, 7, 16, 31, 64, 127]

In [2]: import numpy as np

In [3]: b = np.array(a)

In [4]: b
Out[4]: array([ 0, 0, 1, 4, 7, 16, 31, 64, 127])

In [5]: c = np.array([1, 4., -2, 7])

In [6]: c
Out[6]: array([ 1., 4., -2., 7.])
```

Notice that b is an integer array, as it was created from a list of integers. On the other hand, c is a floating point array even though only one of the elements of the list from which it was made was a floating point number. The array function automatically promotes all of the numbers to the type of the most general entry in the list, which in this case is a floating point number. In the case that elements of a list are made up of numbers and strings, all the elements become strings when an array is formed from a list.

The **second** way arrays can be created is using the NumPy linspace or logspace functions. The linspace function creates an array of N evenly spaced points between a starting point and an ending point. The form of the function is linspace(start, stop, N). If the third argument N is omitted, then N=50.

```
In [7]: np.linspace(0, 10, 5)
Out[7]: array([ 0. , 2.5, 5. , 7.5, 10. ])
```

The `linspace` function produced 5 evenly spaced points between 0 and 10 inclusive. NumPy also has a closely related function `logspace` that produces evenly spaced points on a logarithmically spaced scale. The arguments are the same as those for `linspace` except that `start` and `stop` refer to a power of 10. That is, the array starts at 10^{start} and ends at 10^{stop}.

```
In [8]: %precision 1  # display 1 digit after decimal
Out[8]: '%.1f'
```

```
In [9]: np.logspace(1, 3, 5)
Out[9]: array([ 10. , 31.6, 100. , 316.2, 1000. ])
```

The `logspace` function created an array with 5 points evenly spaced on a logarithmic axis starting at 10^1 and ending at 10^3. The `logspace` function is particularly useful when you want to create a log-log plot.

The **third** way arrays can be created is using the NumPy `arange` function. The form of the function is `arange(start, stop, step)`. If the third argument is omitted `step=1`. If the first and third arguments are omitted, then `start=0` and `step=1`.

```
In [10]: np.arange(0, 10, 2)
Out[10]: array([0, 2, 4, 6, 8])

In [11]: np.arange(0., 10, 2)
Out[11]: array([ 0., 2., 4., 6., 8.])

In [12]: np.arange(0, 10, 1.5)
Out[12]: array([ 0. , 1.5, 3. , 4.5, 6. , 7.5, 9. ])
```

The `arange` function produces points evenly spaced between 0 and 10 exclusive of the final point. Notice that `arange` produces an integer array in the first case but a floating point array in the other two cases. In general `arange` produces an integer array if the arguments are all integers; making any one of the arguments a float causes the array that is created to be a float.

A **fourth** way to create an array is with the `zeros` and `ones` functions. As their names imply, they create arrays where all the elements are either zeros or ones. They each take one mandatory argument, the number of elements in the array, and one optional argument that specifies the data type of the array. Left unspecified, the data type is a float. Here are three examples

Strings, Lists, Arrays, and Dictionaries

```
In [13]: np.zeros(6)
Out[13]: array([ 0., 0., 0., 0., 0., 0.])

In [14]: np.ones(8)
Out[14]: array([ 1., 1., 1., 1., 1., 1., 1., 1.])

In [15]: ones(8, dtype=int)
Out[15]: np.array([1, 1, 1, 1, 1, 1, 1, 1])
```

Recap of ways to create a 1-d NumPy array

array(a): Creates an array from the list a.

linspace(start, stop, num): Returns num evenly spaced numbers over an interval from start to stop inclusive. (num=50 if omitted.)

logspace(start, stop, num): Returns num logarithmically spaced numbers over an interval from 10^{start} to 10^{stop} inclusive. (num=50 if omitted.)

arange([start,] stop[, step,], dtype=None): Returns data points from start to end, exclusive, evenly spaced by step. (step=1 if omitted. start=0 and step=1 if both are omitted.)

zeros(num, dtype=float): Returns an an array of 0s with num elements. Optional dtype argument can be used to set the data type; left unspecified, a float array is made.

ones(num, dtype=float): Returns an an array of 1s with num elements. Optional dtype argument can be used to set the data type; left unspecified, a float array is made.

3.3.2 Mathematical operations with arrays

The utility and power of arrays in Python comes from the fact that you can process and transform all the elements of an array in one fell swoop. The best way to see how this works is to look at an example.

```
In [16]: a = np.linspace(-1., 5, 7)

In [17]: a
Out[17]: array([-1., 0., 1., 2., 3., 4., 5.])

In [18]: a*6
Out[18]: array([ -6., 0., 6., 12., 18., 24., 30.])
```

Here we can see that each element of the array has been multiplied by 6. This works not only for multiplication, but for any other mathematical operation you can imagine: division, exponentiation, *etc.*

```
In [19]: a/5
Out[18]: array([-0.2, 0. , 0.2, 0.4, 0.6, 0.8, 1. ])

In [20]: a**3
Out[20]: array([ -1., 0., 1., 8., 27., 64., 125.])

In [21]: a+4
Out[21]: array([ 3., 4., 5., 6., 7., 8., 9.])

In [22]: a-10
Out[22]: array([-11., -10., -9., -8., -7., -6., -5.])

In [23]: (a+3)*2
Out[23]: array([ 4., 6., 8., 10., 12., 14., 16.])

In [24]: np.sin(a)
Out[24]: array([-0.8415, 0.    ,  0.8415, 0.9093, 0.1411,
                -0.7568, -0.9589])
```

Here we have set `precision 4` so that only 4 digits are displayed to the right of the decimal point. We will typically do this in this manual without mentioning it in order to have neater formatting. Whether or not you do it is entirely up to you.

```
In [25]: np.exp(-a)
Out[25]: array([ 2.7183, 1.    , 0.3679, 0.1353, 0.0498,
                 0.0183, 0.0067])

In [26]: 1. + np.exp(-a)
Out[26]: array([ 3.7183, 2.    , 1.3679, 1.1353, 1.0498,
                 1.0183, 1.0067])

In [27]: b = 5*np.ones(8)

In [28]: b
Out[28]: array([ 5., 5., 5., 5., 5., 5., 5., 5.])

In [29]: b += 4

In [30]: b
Out[30]: array([ 9., 9., 9., 9., 9., 9., 9., 9.])
```

In each case, you can see that the same mathematical operations are performed individually on each element of each array. Even fairly complex algebraic computations can be carried out this way.

Strings, Lists, Arrays, and Dictionaries

Let's say you want to create an $x-y$ data set of $y = \cos x$ vs. x over the interval from -3.14 to 3.14. Here is how you might do it.

```
In [31]: x = np.linspace(-3.14, 3.14, 21)

In [32]: y = np.cos(x)

In [33]: x
Out[33]: array([-3.14 , -2.826, -2.512, -2.198, -1.884,
                -1.57 , -1.256, -0.942, -0.628, -0.314,
                 0.   ,  0.314,  0.628,  0.942,  1.256,
                 1.57 ,  1.884,  2.198,  2.512,  2.826,
                 3.14 ])

In [34]: y
Out[34]: array([ -1.0000e+00, -9.5061e-01, -8.0827e-01,
                 -5.8688e-01, -3.0811e-01,  7.9633e-04,
                  3.0962e-01,  5.8817e-01,  8.0920e-01,
                  9.5111e-01,  1.0000e+00,  9.5111e-01,
                  8.0920e-01,  5.8817e-01,  3.0962e-01,
                  7.9633e-04, -3.0811e-01, -5.8688e-01,
                 -8.0827e-01, -9.5061e-01, -1.0000e+00])
```

You can use arrays as inputs for any of the functions introduced in §2.8.1.

You might well wonder what happens if Python encounters an illegal operation. Here is one example.

```
In [35]: a
Out[35]: array([-1., 0., 1., 2., 3., 4., 5.])

In [36]: np.log(a)
Out[36]: array([    nan,    -inf,     0.   ,  0.6931,  1.0986,
                 1.3863,  1.6094])
```

We see that NumPy calculates the logarithm where it can, and returns nan (not a number) for an illegal operation, taking the logarithm of a negative number, and -inf, or $-\infty$ for the logarithm of zero. The other values in the array are correctly reported. Depending on the settings of your version of Python, NumPy may also print a warning message to let you know that something untoward has occurred.

Arrays can also be added, subtracted, multiplied, and divided by each other on an element-by-element basis, provided the two arrays have the same size. Consider adding the two arrays a and b defined below:

```
In [37]: a = np.array([34., -12, 5.])
```

```
In [38]: b = np.array([68., 5.0, 20.])

In [39]: a+b
Out[39]: array([ 102., -7., 25.])
```

The result is that each element of the two arrays are added. Similar results are obtained for subtraction, multiplication, and division:

```
In [40]: a-b
Out[40]: array([-34., -17., -15.])

In [41]: a*b
Out[41]: array([ 2312., -60., 100.])

In [42]: a/b
Out[42]: array([ 0.5 , -2.4 , 0.25])
```

These kinds of operations with arrays are called *vectorized* operations because the entire array, or "vector," is processed as a unit. Vectorized operations are much faster than processing each element of an array one by one. Writing code that takes advantage of these kinds of vectorized operations is almost always preferred to other means of accomplishing the same task, both because it is faster and because it is usually syntactically simpler. You will see examples of this later on when we discuss loops in Chapter 5.

3.3.3 Slicing and addressing arrays

Arrays can be sliced in the same ways that strings and lists can be sliced—any way you slice it! Ditto for accessing individual array elements: 1-d arrays are addressed the same way as strings and lists. Slicing, combined with the vectorized operations can lead to some pretty compact and powerful code.

Suppose, for example, that we have two arrays y, and t for position *vs.* time of a falling object, say a ball, and we want to use these data to calculate the velocity as a function of time:

```
In [43]: y = np.array([0., 1.3, 5. , 10.9, 18.9, 28.7, 40.])

In [44]: t = np.array([0., 0.49, 1. , 1.5 , 2.08, 2.55, 3.2])
```

We can get find the average velocity for time interval i by the formula

$$v_1 = \frac{y_i - y_{i-1}}{t_i - t_{i-1}}$$

Strings, Lists, Arrays, and Dictionaries

We can easily calculate the entire array of velocities using the slicing and vectorized subtraction properties of NumPy arrays by noting that we can create two y arrays displaced by one index.

```
In [45]: y[:-1]
Out[45]: array([ 0. ,  1.3,  5. , 10.9, 18.9, 28.7])

In [46]: y[1:]
Out[46]: array([ 1.3,  5. , 10.9, 18.9, 28.7, 40. ])
```

The element-by-element difference of these two arrays is

```
In [47]: y[1:]-y[:-1]
Out[47]: array([ 1.3, 3.7, 5.9, 8. , 9.8, 11.3])
```

The element-by-element difference of the two arrays `y[1:]-y[:-1]` divided by `t[1:]-t[:-1]` gives the entire array of velocities.

```
In [48]: v = (y[1:]-y[:-1])/(t[1:]-t[:-1])

In [49]: v
Out[49]: array([  2.6531,   7.2549,  11.8   ,
                 13.7931,  20.8511,  17.3846])
```

Of course, these are the average velocities over each interval so the times best associated with each interval are the times halfway in between the original time array, which we can calculate using a similar trick of slicing:

```
In [50]: tv = (t[1:]+t[:-1])/2.

In [51]: tv
Out[51]: array([ 0.245, 0.745, 1.25 , 1.79 , 2.315, 2.875])
```

3.3.4 Fancy indexing: Boolean masks

There is another way of accessing various elements of an array that is both powerful and useful. We will illustrate with a simple example. Consider the following array

```
In [52]: b = 1.0/np.arange(0.2, 3, 0.2)

In [53]: b
Out[53]:
array([ 5.        ,  2.5       ,  1.66666667,  1.25      ,
        1.        ,  0.83333333,  0.71428571,  0.625     ,
        0.55555556,  0.5       ,  0.45454545,  0.41666667,
        0.38461538,  0.35714286])
```

Suppose we want just those elements of the array that are greater than one. We can get an array of those values using *Boolean indexing*. Here's how it works:

```
In [54]: b[b > 1]
Out[54]:
array([ 5.        ,  2.5       ,  1.66666667,  1.25      ])
```

Only those elements whose values meet the Boolean criterion are returned.

Boolean indexing can be really useful for reassigning values of an array that meet some criterion. For example, we can reassign all the elements of b that are greater than 1 to have a value of 1 with the following assignment:

```
In [55]: b[b > 1] = 1

In [56]: b
Out[56]:
array([ 1.        ,  1.        ,  1.        ,  1.        ,
        1.        ,  0.83333333,  0.71428571,  0.625     ,
        0.55555556,  0.5       ,  0.45454545,  0.41666667,
        0.38461538,  0.35714286])
```

Suppose we create another array that has the same size as b.

```
In [57]: b.size
Out[57]: 14

In [58]: c = np.linspace(0, 10, b.size)

In [59]: c
Out[59]:
array([ 0.        ,  0.76923077,  1.53846154,  2.30769231,
        3.07692308,  3.84615385,  4.61538462,  5.38461538,
        6.15384615,  6.92307692,  7.69230769,  8.46153846,
        9.23076923, 10.        ])
```

Now we would like for this new array c to be equal to 3 everywhere that b is equal to 1. We do that like this:

```
In [60]: c[b == 1] = 3

In [61]: c
Out[61]:
array([ 3.        ,  3.        ,  3.        ,  3.        ,
        3.        ,  3.84615385,  4.61538462,  5.38461538,
        6.15384615,  6.92307692,  7.69230769,  8.46153846,
        9.23076923, 10.        ])
```

Strings, Lists, Arrays, and Dictionaries

Here we have used the Boolean operator `==` which returns a value of `True` if the two things it's comparing have the same value and `False` if they do not. So a Boolean condition on one array can be used to index a different array *if the two arrays have the same size*, as in the above example.

The elements of the array that are selected using Boolean indexing need not be consecutive, as illustrated in the next example.

```
In [62]: y = np.sin(np.linspace(0, 4*np.pi, 9))

In [63]: y
Out[63]:
array([  0.00000000e+00,   1.00000000e+00,   1.22464680e-16,
        -1.00000000e+00,  -2.44929360e-16,   1.00000000e+00,
         3.67394040e-16,  -1.00000000e+00,  -4.89858720e-16])

In [64]: y[np.abs(y) < 1.e-15] = 0

In [65]: y
Out[65]: array([ 0.,  1.,  0., -1.,  0.,  1.,  0., -1.,  0.])
```

Boolean indexing provides a nifty way to get rid of those very small numbers that should be but aren't quite zero due to roundoff error, even if the benefit is mostly aesthetic.

3.3.5 Multi-dimensional arrays and matrices

So far we have examined only one-dimensional NumPy arrays, that is, arrays that consist of a simple sequence of numbers. However, NumPy arrays can be used to represent multidimensional arrays. For example, you may be familiar with the concept of a *matrix*, which consists of a series of rows and columns of numbers. Matrices can be represented using two-dimensional NumPy arrays. Higher dimension arrays can also be created as the application demands.

Creating NumPy arrays

There are a number of ways of creating multidimensional NumPy arrays. The most straightforward way is to convert a list to an array using NumPy's `array` function, which we demonstrate here:

```
In [66]: b = np.array([[1., 4, 5], [9, 7, 4]])

In [67]: b
Out[67]: array([[ 1.,  4.,  5.],
                [ 9.,  7.,  4.]])
```

Notice the syntax used above in which two one-dimensional lists [1., 4, 5] and [9, 7, 4] are enclosed in square brackets to make a two-dimensional list. The array function converts the two-dimensional list, a structure we introduced earlier, to a two-dimensional array. When it makes the conversion from a list to an array, the array function makes all the elements have the same data type as the most complex entry, in this case a float. This reminds us again of an important difference between NumPy arrays and lists: all elements of a NumPy array must be of the same data type: floats, or integers, or complex numbers, *etc*.

There are a number of other functions for creating arrays. For example, a 3 row by 4 column array or 3 × 4 array with all the elements filled with 1 can be created using the ones function introduced earlier.

```
In [68]: a = np.ones((3,4), dtype=float)

In [69]: a
Out[69]: array([[1., 1., 1., 1.],
                [1., 1., 1., 1.],
                [1., 1., 1., 1.]])
```

Using a tuple to specify the size of the array in the first argument of the ones function creates a multidimensional array, in this case a two-dimensional array with the two elements of the tuple specifying the number of rows and columns, respectively. The zeros function can be used in the same way to create a matrix or other multidimensional array of zeros.

The eye(N) function creates an $N \times N$ two-dimensional identity matrix with ones along the diagonal:

```
In [70]: np.eye(4)
Out[70]: array([[1., 0., 0., 0.],
                [0., 1., 0., 0.],
                [0., 0., 1., 0.],
                [0., 0., 0., 1.]])
```

Multidimensional arrays can also be created from one-dimensional arrays using the reshape function. For example, a 2 × 3 array can be created as follows:

```
In [71]: c = np.arange(6)

In [72]: c
Out[72]: array([0, 1, 2, 3, 4, 5])

In [73]: c = np.reshape(c, (2, 3))
```

Strings, Lists, Arrays, and Dictionaries

```
In [74]: c
Out[74]: array([[0, 1, 2],
                [3, 4, 5]])
```

Indexing multidimensional arrays

The individual elements of arrays can be accessed in the same way as for lists:

```
In [75]: b[0][2]
Out[75]: 5.0
```

You can also use the syntax

```
In [76]: b[0, 2]
Out[76]: 5.0
```

which means the same thing. Caution: both the `b[0][2]` and the `b[0, 2]` syntax work for NumPy arrays and mean the same thing; for lists, only the `b[0][2]` syntax works.

Matrix operations

Addition, subtraction, multiplication, division, and exponentiation all work with multidimensional arrays the same way they work with one-dimensional arrays, on an element-by-element basis, as illustrated below:

```
In [77]: b
Out[77]: array([[ 1., 4., 5.],
                [ 9., 7., 4.]])

In [78]: 2*b
Out[78]: array([[ 2., 8., 10.],
                [ 18., 14., 8.]])
In [79]: b/4.
Out[79]: array([[ 0.25, 1. , 1.25],
                [ 2.25, 1.75, 1. ]])

In [80]: b**2
Out[80]: array([[ 1., 16., 25.],
                [ 81., 49., 16.]])

In [81]: b-2
Out[81]: array([[-1., 2., 3.],
                [ 7., 5., 2.]])
```

Functions also act on an element-to-element basis.

```
In [82]: np.sin(b)
Out[82]: array([[ 0.8415, -0.7568, -0.9589],
                [ 0.4121,  0.657 , -0.7568]])
```

Multiplying two arrays together is done on an element-by-element basis. Using the matrices b and c defined above, multiplying them together gives

```
In [83]: b
Out[83]: array([[ 1., 4., 5.],
                [ 9., 7., 4.]])

In [84]: c
Out[84]: array([[0, 1, 2],
                [3, 4, 5]])

In [85]: b*c
Out[85]: array([[ 0.,  4., 10.],
                [ 27., 28., 20.]])
```

Of course, this requires that both arrays have the same shape. Beware: array multiplication, done on an element-by-element basis, is not the same as matrix multiplication as defined in linear algebra. Therefore, we distinguish between *array* multiplication and *matrix* multiplication in Python.

Normal matrix multiplication is done with NumPy's dot function. For example, defining d as:

```
In [86]: d = np.array([[4, 2], [9, 8], [-3, 6]])

In [87]: d
Out[87]: array([[ 4, 2],
                [ 9, 8],
                [-3, 6]])

In [88]: np.dot(b, d)
Out[88]: array([[25., 64.],
                [87., 98.]])
```

3.3.6 Differences between lists and arrays

While lists and arrays are superficially similar—they are both multi-element data structures—they behave quite differently in a number of circumstances. First of all, lists are part of the core Python programming language; arrays are a part of the numerical computing package NumPy. Therefore, you have access to NumPy arrays only if you load the NumPy package using the `import` command.

Here we list some of the differences between Python lists and NumPy arrays, and why you might prefer to use one or the other depending on the circumstance.

- **The elements of a NumPy array must all be of the same type**, whereas the elements of a Python list can be of completely different types.

- **Arrays allow Boolean indexing; lists do not.** See §3.3.4.

- **NumPy arrays support "vectorized" operations** like element-by-element addition and multiplication. This is made possible, in part, by the fact that all elements of the array have the same type, which allows array operations like element-by-element addition and multiplication to be carried out very efficiently (by C loops). Such "vectorized" operations on arrays, which includes operations by NumPy functions such as numpy.sin and numpy.exp, are much faster than operations performed by loops using the core Python math package functions, such as math.sin and math.exp, that act only on individual elements and not on whole lists or arrays.

- **Adding one or more additional elements to a NumPy array creates a new array and destroys the old one.** Therefore it can be very inefficient to build up large arrays by appending elements one by one, especially if the array is very large, because you repeatedly create and destroy large arrays. By contrast, elements can be added to a list without creating a whole new list. If you need to build an array element by element, it is usually better to build it as a list, and then convert it to an array when the list is complete. At this point, it may be difficult for you to appreciate how and under what circumstances you might want build up an array element by element. Examples are provided later on (*e.g.*, see §7.1.1).

3.4 Dictionaries

A Python *list* is a collection of Python objects indexed by an ordered sequence of integers starting from zero. A **dictionary** is also a collection of Python objects, just like a list, but one that is indexed by strings or numbers (not necessarily integers and not in any particular order) or even tuples! Dictionaries are a part of core Python, just like lists.

Suppose we want to make a dictionary of room numbers indexed by the name of the person who occupies each room. We create our dictionary using curly brackets {...}.

```
In [1]: room = {"Emma":309, "Jake":582, "Olivia":764}
```

The dictionary above has three entries separated by commas, each entry consisting of a **key**, which in this case is a string, and a **value**, which in this case is a room number. Each key and its value are separated by a colon. The syntax for accessing the various entries is similar to a that of a list, with the key replacing the index number. For example, to find out the room number of Olivia, we type

```
In [2]: room["Olivia"]
Out[2]: 764
```

The key need not be a string; it can be any immutable Python object. So a key can be a string, an integer, or even a tuple, but it can't be a list. And the elements accessed by their keys need not be a string, but can be almost any legitimate Python object, just as for lists. Here is a weird example.

```
In [3]: weird = {"tank":52, 846:"horse",
   ...: 'bones': [23, 'fox', 'grass'], 'phrase': 'I am here'}

In [4]: weird["tank"]
Out[4]: 52

In [5]: weird[846]
Out[5]: 'horse'

In [6]: weird["bones"]
Out[6]: [23, 'fox', 'grass']

In [7]: weird["phrase"]
Out[7]: 'I am here'
```

Dictionaries can be built up and added to in a straightforward manner

```
In [8]: d = {}

In [9]: d["last name"] = "Alberts"

In [10]: d["first name"] = "Marie"

In [11]: d["birthday"] = "January 27"

In [12]: d
Out[12]: {'birthday': 'January 27', 'first name':
         'Marie', 'last name': 'Alberts'}
```

Strings, Lists, Arrays, and Dictionaries

You can get a list of all the keys or values of a dictionary by typing the dictionary name followed by .keys() or .values().

```
In [13]: d.keys()
Out[13]: ['last name', 'first name', 'birthday']

In [14]: d.values()
Out[14]: ['Alberts', 'Marie', 'January 27']
```

In other languages, data types similar to Python dictionaries may be called "hashmaps" or "associative arrays," so you may see such terms used if you read about dictionaries on the web.

You can also create a dictionary from a list of tuple pairs.

```
In [15]: g = [("Melissa", "Canada"), ("Jeana", "China"),
              ("Etienne", "France")]

In [16]: gd = dict(g)

In [17]: gd
Out[17]: {'Melissa': 'Canada', 'Jeana': 'China',
          'Etienne': 'France'}

In [18]: gd['Jeana']
Out[18]: 'China'
```

3.5 Objects

You may have heard it said that Python is an object-oriented programming language. In fact, we mentioned it on page 2. What it means for a programming language to be object oriented is multi-faceted and involves software and programming design principles that go beyond what we need right now. So rather than attempt some definition that encompasses all of what an object-oriented (OO) approach means, we introduce various aspects of the OO approach that we need as we go along. In this section, our purpose is just to introduce you to some simple (perhaps deceptively so) ideas of the OO approach.

The first of those is the idea of an *object*. One way to think of an object is as a collection of data along with functions that can operate on that data. Virtually everything you encounter in Python is an object, including the data structures we've discussed in this chapter: strings, lists, arrays, and dictionaries. In this case, the data are the contents of these various objects. Associated with each of these kinds of objects are *methods* that act on these objects. For example, consider the string

In [1]: c = "My dog's name is Bingo"

One method associated with a string object is `split()`, which we invoke using the dot notation we've encountered before:

In [2]: c.split()
Out [2]: ['My', "dog's", 'name', 'is', 'Bingo']

The method `split()` acts on the object it's attached to by the dot. It has a matching set of parentheses to indicate that it's a function (that acts on the object a). Without an argument, `split()` splits the string a at the spaces into separate strings, five in this case, and returns the set of split strings as a list. By specifying an argument to `split()`, we can split the string elsewhere, say at the "g":

In [3]: c.split('g')
Out [3]: ['My do', "'s name is Bin", 'o']

There are many more string methods, which we do not explore here, as our point isn't to give an exhaustive introduction to string methods and their uses. Instead, it's simply to introduce the idea of object methods, and to start introducing Python objects in a concrete way.

Lists, arrays, and dictionaries are also objects that have methods associated with them. Consider the following 2-row array:

In [4]: b = np.array([[1., 4., 5.], [9., 7., 4.]])

In [5]: b
Out[5]: array([[1., 4., 5.],
               [9., 7., 4.]])
In [6]: b.mean()
Out[6]: 5.0

In [7]: b.shape
Out[7]: (2, 3)

The method `mean()` calculates the mean value of the elements of the array. Writing `b.shape` returns the number of rows and columns in the array. Note, that there are no parentheses associated with `b.shape`. That's because `b.shape` contains *data* associated with the array object b; it's not calculated by a method, but is in fact simply stored in the object. Writing `b.shape` simply looks up the data within the object and reports it back to you. Data stored like this with an object is called an *instance variable*. We'll have more to say about instance variables later on (and why the word "instance" is used) but for now it's sufficient just to learn the jargon.

In general, objects have *methods*, which are functions that act on

Strings, Lists, Arrays, and Dictionaries

the object, and *instance variables*, which are data stored with the object, associated with them. Taken together, methods and instance variables are known as *attributes* of an object. Each kind of object, for example, strings, lists, arrays, and dictionaries, has its own set of attributes, which are uniquely associated with that object class. So while `split()` is an attribute associated with strings, it is not an attribute of arrays. Thus, typing `b.split()` returns an error message, since `b` is an array.

```
In [8]: b.split()
Traceback (most recent call last):

  File "<ipython-input-11-0c30fe27ab6f>", line 1, in <module>
    b.split()

AttributeError: 'numpy.ndarray' object has no attribute 'split'
```

We will have many occasions to exploit object attributes in our Python journey. The table below summarize some of the attributes of NumPy arrays.

Instance variable	Output
.size	number of elements in array
.shape	number of rows, columns, *etc.*
.ndim	number of array dimensions
.real	real part of array
.imag	imaginary part of array
method	**Output**
.mean()	average value of array elements
.std()	standard deviation of array elements
.min()	return minimum value of array
.max()	return maximum value of array
.sort()	low-to-high sorted array (in place)
.reshape(a, b)	Returns an a×b array with same elements
.conj()	complex-conjugate all elements

Table 3.1 NumPy array attributes.

3.6 Exercises

1. Create an array of 9 evenly spaced numbers going from 0 to 29 (inclusive) and give it the variable name r. Find the square of each

element of the array (as simply as possible). Find twice the value of each element of the array in two different ways: (i) using addition and (ii) using multiplication. Print out the array r and each output requested above.

2. Create the following arrays:

   (a) an array of 100 elements all equal to e, the base of the natural logarithm;

   (b) an array in 1-degree increments of all the angles in degrees from 0 to 360 degrees inclusive;

   (c) an array in 1-degree increments of all the angles in radians from 0 to 360 degrees inclusive. Verify your answers by showing that `c-b*np.pi/180` gives an array of zeros (or nearly zeros) where b and c are the arrays you created in parts (b) and (c);

   (d) an array from 12 to 17, not including 17, in 0.2 increments;

   (e) an array from 12 to 17, including 17, in 0.2 increments.

3. The position of a ball at time t dropped with zero initial velocity from a height h_0 is given by

$$y = h_0 - \frac{1}{2}gt^2$$

where $g = 9.8$ m/s^2. Suppose $h_0 = 10$ m. Find the sequence of times when the ball passes each half meter assuming the ball is dropped at $t = 0$. Hint: Create a NumPy array for y that goes from 10 to 0 in increments of -0.5 using the `arange` function. Solving the above equation for t, show that

$$t = \sqrt{\frac{2(h_0 - y)}{g}}.$$

Using this equation and the array you created, find the sequence of times when the ball passes each half meter. Save your code as a Python script. It should yield the following results for the y and t arrays:

```
In [1]: y
Out[1]: array([10. ,  9.5,  9. ,  8.5,  8. ,
                7.5,  7. ,  6.5,  6. ,  5.5,  5. ,
```

```
                4.5, 4. , 3.5, 3. , 2.5, 2. ,
                1.5, 1. , 0.5])

In [2]: t
Out[2]: array([ 0.   ,  0.31943828,  0.45175395,
        0.55328334,  0.63887656,  0.71428571,
        0.7824608 ,  0.84515425,  0.9035079 ,
        0.95831485,  1.01015254,  1.05945693,
        1.10656667,  1.15175111,  1.19522861,
        1.23717915,  1.27775313,  1.31707778,
        1.35526185,  1.39239919])
```

Once you have created the arrays y and t, type `list(zip(t, y))` at the IPython prompt. Explain the result.

4. Recalling that the average velocity over an interval Δt is defined as $\bar{v} = \Delta y/\Delta t$, find the average velocity for each time interval in the previous problem using NumPy arrays. Keep in mind that the number of time intervals is one less than the number of times. Hint: What are the arrays `y[1:20]` and `y[0:19]`? What does the array `y[1:20]-y[0:19]` represent? (Try printing out the two arrays from the IPython shell.) Using this last array and a similar one involving time, find the array of average velocities. Bonus: Can you think of a more elegant way of representing `y[1:20]-y[0:19]` that does not make explicit reference to the number of elements in the y array—one that would work for any length array?

You should get the following answer for the array of velocities:

```
In [5]: v
Out[5]: array([ -1.56524758,  -3.77884195,
        -4.9246827 ,  -5.84158351,  -6.63049517,
        -7.3340579 ,  -7.97531375,  -8.56844457,
        -9.12293148,  -9.64549022, -10.14108641,
       -10.61351563, -11.06575711, -11.50020061,
       -11.91879801, -12.32316816, -12.71467146,
       -13.09446421, -13.46353913])
```

BONUS: Calculate the acceleration as a function of time using the formula $\bar{a} = \Delta v/\Delta t$. Take care, as you will need to define a new time array that corresponds to the times where the velocities are calculated, which is midway between the times in the original time array. You should be able to justify the answer you get for the array of accelerations.

5. Perform the following tasks with NumPy arrays. All of them can be done (elegantly) in 1 to 3 lines.

   (a) Create an 8 × 8 array with ones on all the edges and zeros everywhere else.

   (b) Create an 8 × 8 array of integers with a checkerboard pattern of ones and zeros.

   (c) Given the array `c = np.arange(2, 50, 5)`, make all the numbers not divisible by 3 negative.

   (d) Find the size, shape, mean, and standard deviation of the arrays you created in parts (a)–(c).

CHAPTER 4

Input and Output

> *In this chapter, you learn how to **input or read data** into a Python program, either from the keyboard or a computer file. You also learn how to **output or write data**, either to a computer screen or to a computer file.*

A good relationship depends on good communications. In this chapter you learn how to communicate with Python. Of course, communicating is a two-way street: input and output. Generally, when you have Python perform some task, you need to feed it information—input. When it is done with that task, it reports back to you the results of its calculations—output.

There are two venues for input that concern us: the computer keyboard and the input data file. Similarly, there are two venues for output: the computer screen and the output data file. We start with input from the keyboard and output to the computer screen. Then we deal with data file input and output—or "io."

4.1 Keyboard Input

Many computer programs need input from the user. In §2.7.1, the program `myTrip.py` required the distance traveled as an input in order to determine the duration of the trip and the cost of the gasoline. As you might like to use this same script to determine the cost of several different trips, it would be useful if the program requested that input when it was run from the IPython shell.

Python has a function called `input` for getting input from the user and assigning it a variable name. It has the form

```
strname = input("prompt to user")
```

When the `input` function is executed, it prints to the computer screen the text in the quotes and waits for input from the user. The user types a string of characters and presses the return key. The `input` function then assigns that string to the variable name on the right of the assignment operator =. Let's try it out with this snippet of code in the IPython shell.

```
In [1]: distance = input("Input trip distance (miles): ")
Input trip distance (miles):
```

Python prints out the string argument of the `input` function and waits for a response from you. Let's go ahead and type 450 for "450 miles" and press return. Now type the variable name `distance` to see its value.

```
In [2]: distance
Out[2]: '450'
```

The value of the `distance` is 450 as expected, but it is a string, as you can see, because 450 is enclosed in quotes. Because we want to use 450 as a number and not a distance, we need to convert it from a string to a number. We can do that with the `eval` function by writing

```
In [3]: distance = eval(distance)

In [4]: distance
Out[4]: 450
```

The `eval` function has converted *distance* to an integer. This is fine and we are ready to move on. However, we might prefer that `distance` be a float instead of an integer. There are two ways to do this. We could assume the user is very smart and will type "450." instead of "450", which will cause distance to be a float when `eval` does the conversion. That is, the number 450 is dynamically typed to be a float or an integer depending on whether or not the user uses a decimal point. Alternatively, we could use the function `float` in place of `eval`, which would ensure that `distance` is a floating point variable. Thus, our code would look like this (including the user response):

```
In [5]: distance = input("Input distance of trip (miles): ")
Input distance of trip (miles): 450

In [5]: distance
Out[5]: '450'
```

Input and Output

```
In [7]: distance = float(distance)

In [8]: distance
Out[8]: 450.0
```

Now let's incorporate what we have learned into the code we wrote as our first scripting example in §2.7.1

Code: chapter4/programs/myTripIO.py

```
1  """Calculates time, gallons of gas used, and cost of gasoline
2  for a trip"""
3
4  distance = input("Input trip distance (miles): ")
5  distance = float(distance)
6
7  mpg = 30.              # car mileage
8  speed = 60.            # average speed
9  costPerGallon = 2.85   # price of gasoline
10
11 time = distance/speed
12 gallons = distance/mpg
13 cost = gallons*costPerGallon
14 print(time, gallons, cost)
```

Note that we have put our comment explaining what this script does between triple quotes distributed over two lines. Comments like this, between triple quotes (single or double) are called *docstrings*. Everything between the triple quotes is part of the docstring, which can extend over multiple lines, as it does here. It's a good idea to include a docstring explaining what your script does at the beginning of your file.

Lines 4 and 5 can be combined into a single line, which is a little more efficient:

```
distance = float(input("Input trip distance (miles): "))
```

Whether you use `float` or `int` or `eval` depends on whether you want a float, an integer, or a dynamically typed variable. In this program, it doesn't matter, but in general it's good practice to explicitly cast the variable in the type you would like it to have. Here `distance` is used as a float so it's best to cast it as such, as we do in the example above.

Now you simply run the program and then type `time`, `gallons`, and `cost` to view the results of the calculations done by the program.

Before moving on to output, we note that sometimes you may want string input rather that numerical input. For example, you might want the user to input their name, in which case you would simply use the `input` function without converting its output.

4.2 Screen Output

It would be much more convenient if the program in the previous section would simply write its output to the computer screen, instead of requiring the user to type `time`, `gallons`, and `cost` to view the results. Fortunately, this can be accomplished very simply using Python's `print` function. For example, simply including the statement `print(time, gallons, cost)` after line 13, running the program would give the following result:

```
In [1]: run myTripIO.py
What is the distance of your trip (miles)? 450
7.5 15.0 42.75
```

The program prints out the results as a tuple of time (in hours), gasoline used (in gallons), and cost (in dollars). Of course, the program doesn't give the user a clue as to which quantity is which. The user has to know.

4.2.1 Formatting output with `str.format()`

We can clean up the output of the example above and make it considerably more user friendly. The program below demonstrates how to do this.

Code: chapter4/programs/myTripNiceIO.py

```
1  """Calculates time, gallons of gas used, and cost of gasoline
2  for a trip"""
3
4  distance = float(input("Input trip distance (miles): "))
5
6  mpg = 30.              # car mileage
7  speed = 60.            # average speed
8  costPerGallon = 2.85   # price of gasoline
9
10 time = distance/speed
11 gallons = distance/mpg
12 cost = gallons*costPerGallon
13
```

Input and Output

```
14  print("\nDuration of trip = {0:0.1f} hours".format(time))
15  print("Gasoline used = {0:0.1f} gallons (@ {1:0.0f} mpg)"
16        .format(gallons, mpg))
17  print("Cost of gasoline = ${0:0.2f} (@ ${1:0.2f}/gallon)"
18        .format(cost, costPerGallon))
```

Running this program, with the distance provided by the user, gives

```
In [9]: run myTripNiceIO.py

Input trip distance (miles): 450

Duration of trip = 7.5 hours
Gasoline used = 15.0 gallons (@ 30 mpg)
Cost of gasoline = $42.75 (@ $2.85/gallon)
```

Now the output is presented in a way that is immediately understandable to the user. Moreover, the numerical output is formatted with an appropriate number of digits to the right of the decimal point. For good measure, we also included the assumed mileage (30 mpg) and the cost of the gasoline. All of this is controlled by the str.format() function within the print function.

The argument of the print function is of the form str.format() where str is a string that contains text that is to be written the screen, as well as certain format specifiers contained within curly braces { }, which we discuss below. The format method (a method of string objects) contains the list of variables that are to be printed.

- The \n at the start of the string in the print statement on line 14 is the newline character. It creates the blank line before the output is printed.

- The positions of the curly braces specify where the variables in the format function at the end of the statement are printed.

- The format string inside the curly braces specifies how each variable in the format function is printed.

- The number before the colon in the format string specifies which variable in the list in the format function is printed. Remember, Python is zero-indexed, so 0 means the first variable is printed, 1 means the second variable, *etc.*

- The zero after the colon specifies the *minimum* number of spaces reserved for printing out the variable in the format function. A zero means that only as many spaces as needed will be used.

- The number after the period specifies the number of digits to the right of the decimal point that will be printed: 1 for `time` and `gallons` and 2 for `cost`.

- The letter `f` specifies that a number is to be printed with a *fixed* number of digits. If the `f` format specifier is replaced with `e`, then the number is printed out in exponential format (scientific notation).

In addition to `f` and `e` format types, there are two more that are commonly used: `d` for integers (digits) and `s` for strings. There are, in fact, many more formatting possibilities. Python has a whole "Format Specification Mini-Language" that is documented at:
http://docs.python.org/library/string.html#formatspec.
It's very flexible but arcane. You might find it simplest to look at the "Format examples" section further down the same web page.

Finally, note that the code starting on lines 15 and 17 are each split into two lines. We have done this so that the lines fit on the page without running off the edge. Python allows you to break lines up that are inside parentheses to improve readability. More information about line continuation in Python can be found here: http://www.python.org/dev/peps/pep-0008/.

The program below illustrates most of the formatting you will need for writing a few variables, be they strings, integers, or floats, to screen or to data files (which we discuss in the next section).

Code: chapter4/programs/printFormatExamples.py

```
1  string1 = "How"
2  string2 = "are you my friend?"
3  int1 = 34
4  int2 = 942885
5  float1 = -3.0
6  float2 = 3.141592653589793e-14
7  print(string1)
8  print(string1 + ' ' + string2)
9  print('A. {} {}'.format(string1, string2))
10 print('B. {0:s} {1:s}'.format(string1, string2))
11 print('C. {0:s} {0:s} {1:s} - {0:s} {1:s}'
12       .format(string1, string2))
13 print('D. {0:10s}{1:5s}'           # reserves 10 & 5 spaces,
```

Input and Output

```
14              .format(string1, string2))    # respectively for 2 strings
15     print(' **')
16     print(int1, int2)
17     print('E. {0:d} {1:d}'.format(int1, int2))
18     print('F. {0:8d} {1:10d}'.format(int1, int2))
19     print(' ***')
20     print('G. {0:0.3f}'.format(float1))    # 3 decimal places
21     print('H. {0:6.3f}'.format(float1))    # 6 spaces, 3 decimals
22     print('I. {0:8.3f}'.format(float1))    # 8 spaces, 3 decimals
23     print(2*'J. {0:8.3f}    '.format(float1))
24     print(' ****')
25     print('K. {0:0.3e}'.format(float2))
26     print('L. {0:12.3e}'.format(float2))   # 12 spaces, 3 decimals
27     print('M. {0:12.3f}'.format(float2))   # 12 spaces, 3 decimals
28     print(' *****')
29     print('N. 12345678901234567890')
30     print('O. {0:s}--{1:8d},{2:10.3e}'
31              .format(string2, int1, float2))
```

Here is the output:

```
How
How are you my friend?
A. How are you my friend?
B. How are you my friend?
C. How How are you my friend? - How are you my friend?
D. How          are you my friend?
 **
34 942885
E. 34 942885
F.       34     942885
 ***
G. -3.000
H. -3.000
I.   -3.000
J.   -3.000    J.   -3.000
 ****
K. 3.142e-14
L.    3.142e-14
M.        0.000
 *****
N. 12345678901234567890
O. are you my friend?--       34, 3.142e-14
```

Successive empty brackets {} like those that appear in line 9 will print out in order the variables appear inside the `format()` method using their default format. Starting with line 10, the number to the left of the colon inside the curly brackets specifies which of the variables, numbered starting with 0, in the `format` method is printed. The characters that appear to the right of the colon are the format specifiers with the following correspondences: s–string, d–integer, f–fixed floating point number, e–exponential floating point number. The format specifiers `6.3f` and `8.3f` in lines 21 and 22 tell the `print` statement to reserve at least 6 and 8 total spaces, respectively, with 3 decimal places for the output of a floating point number. Studying the output of the other lines will help you understand how formatting works.

4.2.2 Printing arrays

Formatting NumPy arrays for printing requires another approach. As an example, let's create an array and then format it in various ways. From the IPython terminal:

```
In [10]: a = np.linspace(3, 19, 7)
In [11]: print(a)
[  3.      5.6667   8.3333  11.     13.6667  16.3333  19.   ]
```

Simply using the `print` function does print out the array, but perhaps not in the format you desire. To control the output format, you use the NumPy function `set_printoptions`. For example, suppose you want to see no more than two digits to the right of the decimal point. Then you simply write

```
In [12]: np.set_printoptions(precision=2)
In [13]: print(a)
[  3.    5.67  8.33 11.   13.67 16.33 19.  ]
```

If you want to change the number of digits to the right of the decimal point to 4, you set the keyword argument `precision` to 4.

```
In [14]: np.set_printoptions(precision=4)
In [15]: print(a)
[  3.      5.6667   8.3333  11.     13.6667  16.3333  19.   ]
```

Suppose you want to use scientific notation. The method for doing it is somewhat arcane, using something called a `lambda` function. For now, you don't need to understand how it works to use it. Just follow the examples shown below, which illustrate several different output formats using the `print` function with NumPy arrays.

Input and Output

```
In [16]: np.set_printoptions(
    ...: formatter={'float': lambda x: format(x, '5.1e')})

In [17]: print(a)
[3.0e+00 5.7e+00 8.3e+00 1.1e+01 1.4e+01 1.6e+01 1.9e+01]
```

To specify the format of the output, you use the `formatter` keyword argument. The first entry to the right of the curly bracket is a string that can be `'float'`, as it is above, or `'int'`, or `'str'`, or a number of other data types that you can look up in the online NumPy documentation. The only other thing you should change is the format specifier string. In the above example, it is `'5.1e'`, specifying that Python should allocate at least 5 spaces, with 1 digit to the right of the decimal point in scientific (exponential) notation. For fixed-width floats with 3 digits to the right of the decimal point, use the `f` in place of the `e` format specifier, as follows:

```
In [18]: np.set_printoptions(
    ...: formatter={'float': lambda x: format(x, '6.3f')})
In [19]: print(a)
[ 3.000  5.667  8.333 11.000 13.667 16.333 19.000]
```

To return to the default format, type the following:

```
In [20]: np.set_printoptions(precision=8)
In [21]: print(a)
[ 3.         5.66666667  8.33333333 11.         13.66666667 16.33333333 19.        ]
```

4.3 File Input

4.3.1 Reading data from a text file

Often you would like to analyze data that you have stored in a text file. Consider, for example, the data file below for an experiment measuring the free fall of a mass.

Data: chapter4/programs/mydata.txt

```
Data for falling mass experiment
Date: 16-Aug-2016
Data taken by Lauren and John

data point   time (sec)   height (mm)   uncertainty (mm)
    0           0.0          180             3.5
    1           0.5          182             4.5
```

2	1.0	178	4.0
3	1.5	165	5.5
4	2.0	160	2.5
5	2.5	148	3.0
6	3.0	136	2.5
7	3.5	120	3.0
8	4.0	99	4.0
9	4.5	83	2.5
10	5.0	55	3.6
11	5.5	35	1.75
12	6.0	5	0.75

We would like to read these data into a Python program, associating the data in each column with an appropriately named array. While there are a multitude of ways to do this in Python, the simplest by far is to use the NumPy `loadtxt` function, whose use we illustrate here. Suppose that the name of the text file is `mydata.txt`. Then we can read the data into four different arrays with the following statement:

```
In [1]: dataPt, time, height, error = np.loadtxt(
                "mydata.txt", skiprows=5 , unpack=True)
```

In this case, the `loadtxt` function takes three arguments: the first is a string that is the name of the file to be read, the second tells `loadtxt` to skip the first 5 lines at the top of file, sometimes called the *header*, and the third tells `loadtxt` to output the data (*unpack* the data) so that it can be directly read into arrays. `loadtxt` reads however many columns of data are present in the text file to the array names listed to the left of the "=" sign. The names labeling the columns in the text file are not used, but you are free to choose the same or similar names, of course, as long as they are legal array names. By the way, for the above `loadtxt` call to work, the file `mydata.txt` should be in the current working directory of the IPython shell. Otherwise, you need to specify the directory path with the file name.

It is critically important that the data file be a *text* file. It cannot be an MS Word file, for example, or an Excel file, or anything other than a plain text file. Such files can be created by text editor programs like **Notepad++** (for PCs) or **BBEdit** (for Macs). They can also be created by MS Word and Excel provided you explicitly save the files as text files. **Beware**: You should exit any text file you make and save it with a program that allows you to save the text file using **UNIX**-type formatting, which uses a *line feed* (LF) to end a line. Some programs, like MS Word under Windows, may include a carriage return (CR) char-

Input and Output

acter, which can confuse `loadtxt`. Note that we give the file name a
`.txt` *extension*, which indicates to most operating systems that this is
a *text* file, as opposed to an Excel file, for example, which might have
a `.xlsx` or `.xls` extension.

If you don't want to read in all the columns of data, you can specify which columns to read in using the `usecols` keyword. For example, the call

```
In [2]: time, height = np.loadtxt('mydata.txt', skiprows=5,
                                  usecols=(1,2), unpack=True)
```

reads in only columns 1 and 2; columns 0 and 3 are skipped. As a consequence, only two array names are included to the left of the "=" sign, corresponding to the two columns that are read. Writing `usecols = (0,2,3)` would skip column 1 and read in only the data in colums 0, 2, and 3. In this case, 3 array names would need to be provided on the left-hand side of the "=" sign.

One convenient feature of the `loadtxt` function is that it recognizes any *white space* as a column separator: spaces, tabs, *etc.*

Finally you should remember that `loadtxt` is a NumPy function. So if you are using it in a Python module, you must be sure to include an "`import numpy as np`" statement before calling "`np.loadtxt`".

4.3.2 Reading data from an Excel file: CSV files

Sometimes you have data stored in a spreadsheet program like Excel that you would like to read into a Python program. The *Excel data sheet* shown in Fig. 4.1 contains the same data set we saw above in a text file. While there are a number of different approaches one can use to read such files, one of the simplest and most robust is to save the spreadsheet as a CSV ("comma-separated value") file, a format which all common spreadsheet programs can create and read. So, if your Excel spreadsheet was called `mydata.xlsx`, the CSV file saved using Excel's Save As command would by default be `mydata.csv`. It would look like this:

Data: chapter4/programs/mydata.csv

```
Data for falling mass experiment,,,
Date: 16-Aug-2016,,,
Data taken by Lauren and John,,,
,,,
data point,time (sec),height (mm),uncertainty (mm)
```

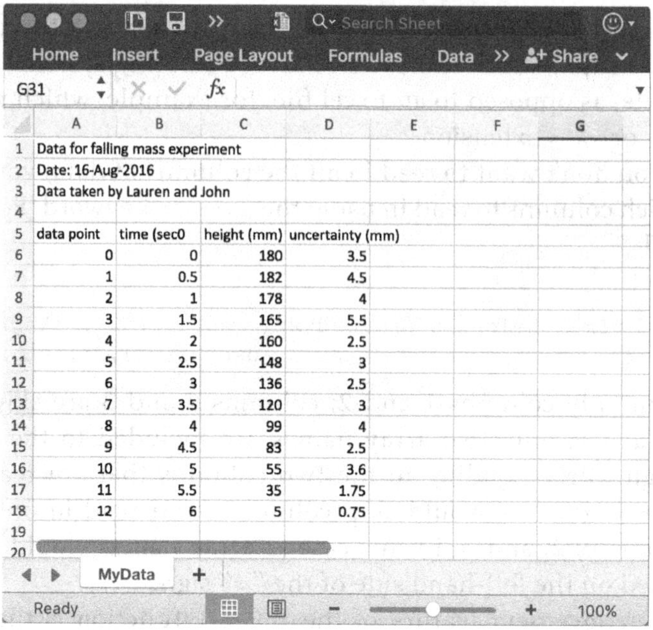

	A	B	C	D	E	F	G
1	Data for falling mass experiment						
2	Date: 16-Aug-2016						
3	Data taken by Lauren and John						
4							
5	data point	time (sec0	height (mm)	uncertainty (mm)			
6	0	0	180	3.5			
7	1	0.5	182	4.5			
8	2	1	178	4			
9	3	1.5	165	5.5			
10	4	2	160	2.5			
11	5	2.5	148	3			
12	6	3	136	2.5			
13	7	3.5	120	3			
14	8	4	99	4			
15	9	4.5	83	2.5			
16	10	5	55	3.6			
17	11	5.5	35	1.75			
18	12	6	5	0.75			

Figure 4.1 Excel data sheet.

```
0,0,180,3.5
1,0.5,182,4.5
2,1,178,4
3,1.5,165,5.5
4,2,160,2.5
5,2.5,148,3
6,3,136,2.5
7,3.5,120,3
8,4,99,4
9,4.5,83,2.5
10,5,55,3.6
11,5.5,35,1.75
12,6,5,0.75
```

As its name suggests, the CSV file is simply a text file with the data that was formerly in spreadsheet columns now separated by commas. We can read the data in this file into a Python program using the `loadtxt` NumPy function once again. Here is the code

```
In [3]: dataPt, time, height, error = np.loadtxt("mydata.csv",
                skiprows=5 , unpack=True, delimiter=',')
```

The form of the function is exactly the same as before except we

Input and Output

have added the argument `delimiter=','` that tells `loadtxt` that the columns are separated by commas instead of white space (spaces or tabs), which is the default. Once again, we set the `skiprows` argument to skip the header at the beginning of the file and to start reading at the first row of data. The data are output to the arrays to the right of the assignment operator = exactly as in the previous example.

4.4 File Output

4.4.1 Writing data to a text file

There is a plethora of ways to write data to a data file in Python. We will stick to one very simple one that's suitable for writing data files in text format. It uses the NumPy `savetxt` routine, which is the counterpart of the `loadtxt` routine introduced in the previous section. The general form of the routine is

```
savetxt(filename, array, fmt="%0.18e",
        delimiter=" ", newline=" \n", header="",
        footer="", comments="#")
```

We illustrate `savetext` below with a script that first creates four arrays by reading in the data file `mydata.txt`, as discussed in the previous section, and then writes that same data set to another file `mydataout.txt`.

Code: chapter4/programs/ReadWriteMyData.py

```
import numpy as np
dataPt, time, height, error = np.loadtxt("mydata.txt",
                                          skiprows=5,
                                          unpack=True)
np.savetxt("mydatawritten.txt",
           list(zip(dataPt, time, height, error)),
           fmt="%12.1f")
```

The first argument of `savetxt` is a string, the name of the data file to be created. Here we have chosen the name `mydataout.txt`, inserted with quotes, which designates it as a string literal. Beware, if there is already a file of that name on your computer, it will be overwritten—the old file will be destroyed and a new one will be created.

The second argument is the data array that is to be written to the data file. Because we want to write not one but four data arrays to the file, we have to package the four data arrays as one, which we do using the `zip` function, a Python function that combines the four arrays and returns a list of tuples, where the ith tuple contains the ith element from each of the arrays (or lists, or tuples) listed as its arguments. Since there are four arrays, each row will be a tuple with four entries, producing a table with four columns. In fact, the `zip` function is just a set of instructions to produce each tuple one after another; the `list` function is needed to actually construct the entire list of tuples.[1] Note that the first two arguments, the `filename` and data `array`, are regular arguments and thus must appear as the first and second arguments in the correct order. The remaining arguments are all *keyword arguments*, meaning that they are optional and can appear in any order, provided you use the keyword.

The next argument is a format string that determines how the elements of the array are displayed in the data file. The argument is optional and, if left out, is the format `0.18e`, which displays numbers as 18 digit floats in exponential (scientific) notation. Here we choose a different format, `12.1f`, which is a float displayed with 1 digit to the right of the decimal point and a minimum width of 12. By choosing 12, which is more digits than any of the numbers in the various arrays have, we ensure that all the columns will have the same width. It also ensures that the decimal points in a column of numbers are aligned. This is evident in the data file below, `mydatawritten.txt`, which was produced by the above script.

Data: chapter4/programs/mydatawritten.txt

```
         0.0         0.0       180.0         3.5
         1.0         0.5       182.0         4.5
         2.0         1.0       178.0         4.0
         3.0         1.5       165.0         5.5
         4.0         2.0       160.0         2.5
         5.0         2.5       148.0         3.0
         6.0         3.0       136.0         2.5
         7.0         3.5       120.0         3.0
         8.0         4.0        99.0         4.0
         9.0         4.5        83.0         2.5
        10.0         5.0        55.0         3.6
        11.0         5.5        35.0         1.8
        12.0         6.0         5.0         0.8
```

---

[1]Technically, the `zip` function is an *iterator*, like the `range` function introduced in §3.2.2. We discuss iterators more fully when we discuss the `range` function in §5.2.

Input and Output

We omitted the optional `delimiter` keyword argument, which leaves the delimiter as the default space. We also omitted the optional `header` keyword argument, which is a string variable that allows you to write header text above the data. For example, you might want to label the data columns and also include the information that was in the header of the original data file. To do so, you just need to create a string with the information you want to include and then use the `header` keyword argument. The code below illustrates how to do this.

Code: chapter4/programs/ReadWriteMyDataHeader.py

```python
import numpy as np

dataPt, time, height, error = np.loadtxt("MyData.txt",
                                         skiprows=5,
                                         unpack=True)

info = 'Data for falling mass experiment'
info += '\nDate: 16-Aug-2016'
info += '\nData taken by Lauren and John'
info += '\n\n data point time (sec) height (mm) '
info += 'uncertainty (mm)'

np.savetxt('ReadWriteMyDataHeader.txt',
           list(zip(dataPt, time, height, error)),
           header=info, fmt="%12.1f")
```

Now the data file produced has a header preceding the data. Notice that the header rows all start with a # comment character, which is the default setting for the `savetxt` function. This can be changed using the keyword argument `comments`. You can find more information about `savetxt` using the IPython `help` function or from the online NumPy documentation.

Data: chapter4/programs/ReadWriteMyDataHeader.txt

```
# Data for falling mass experiment
# Date: 16-Aug-2016
# Data taken by Lauren and John
#
#   data point  time (sec)   height (mm)   uncertainty (mm)
         0.0         0.0         180.0         3.5
         1.0         0.5         182.0         4.5
         2.0         1.0         178.0         4.0
         3.0         1.5         165.0         5.5
         4.0         2.0         160.0         2.5
         5.0         2.5         148.0         3.0
         6.0         3.0         136.0         2.5
```

```
         7.0          3.5         120.0          3.0
         8.0          4.0          99.0          4.0
         9.0          4.5          83.0          2.5
        10.0          5.0          55.0          3.6
        11.0          5.5          35.0          1.8
        12.0          6.0           5.0          0.8
```

4.4.2 Writing data to a CSV file

To produce a CSV file, you would specify a comma as the delimiter. You might use the `0.1f` format specifier, which leaves no extra spaces between the comma data separators, as the file is to be read by a spreadsheet program, which will determine how the numbers are displayed. The code, which could be substituted for the `savetxt` line in the above code, reads

```python
np.savetxt('mydataout.csv',
           list(zip(dataPt, time, height, error)),
           fmt="%0.1f", delimiter=",")
```

and produces the following data file.

Data: chapter4/programs/mydataout.csv

```
0.0,0.0,180.0,3.5
1.0,0.5,182.0,4.5
2.0,1.0,178.0,4.0
3.0,1.5,165.0,5.5
4.0,2.0,160.0,2.5
5.0,2.5,148.0,3.0
6.0,3.0,136.0,2.5
7.0,3.5,120.0,3.0
8.0,4.0,99.0,4.0
9.0,4.5,83.0,2.5
10.0,5.0,55.0,3.6
11.0,5.5,35.0,1.8
12.0,6.0,5.0,0.8
```

This data file, with a `csv` extension, can be directly read by a spreadsheet program like Excel.

4.5 Exercises

1. Write a Python program that calculates how much money you can spend each day for lunch for the rest of the month based on to-

Input and Output

day's date and how much money you currently have in your lunch account. The program should ask you: (1) how much money you have in your account, (2) what today's date is, and (3) how many days there are in the month. The program should return your daily allowance. The results of running your program should look like this:

```
How much money (in dollars) in your lunch account? 319
What day of the month is today? 21
How many days in this month? 30
You can spend $31.90 each day for the rest of the month.
```

Extra: Create a dictionary (see §3.4) that stores the number of days in each month (forget about leap years) and have your program ask what month it is rather than the number of days in the month.

2. From the IPython terminal, create the following three NumPy arrays:

```
a = array([1, 3, 5, 7])
b = array([8, 7, 5, 4])
c = array([0, 9,-6,-8])
```

Now use the `zip` function to create a list d defined as

```
d = list(zip(a,b,c))
```

What kind of object is each of the four elements of the list d? Convert d into a NumPy array and call that array e. Type e at the terminal prompt so that e is printed out on the IPython terminal. One of the elements of e is -8. Show how to address and print out just that element of e. Show how to address that same element of d. What kind of object is e[1]?

3. Create the following data file and then write a Python script to read it into three NumPy arrays with the variable names f, a, da for the frequency, amplitude, and amplitude error.

```
Date: 2013-09-16
Data taken by Liam and Selena
frequency (Hz)   amplitude (mm)   amp error (mm)
      0.7500          13.52              0.32
      1.7885          12.11              0.92
      2.8269          14.27              0.73
      3.8654          16.60              2.06
```

```
 4.9038        22.91        1.75
 5.9423        35.28        0.91
 6.9808        60.99        0.99
 8.0192        33.38        0.36
 9.0577        17.78        2.32
10.0962        10.99        0.21
11.1346         7.47        0.48
12.1731         6.72        0.51
13.2115         4.40        0.58
14.2500         4.07        0.63
```

Show that you have correctly read in the data by having your script print out to your computer screen the three arrays. Format the printing so that it produces output like this:

```
f =
[ 0.75 1.7885 2.8269 3.8654 4.9038 5.9423
6.9808 8.0192 9.0577 10.0962 11.1346 12.1731
13.2115 14.25 ]
a =
[ 13.52 12.11 14.27 16.6 22.91 35.28 60.99
33.38 17.78 10.99 7.47 6.72 4.4 4.07]
da =
[ 0.32 0.92 0.73 2.06 1.75 0.91 0.99 0.36 2.32
0.21 0.48  0.51 0.58 0.63]
```

Note that the array f is displayed with four digits to the right of the decimal point while the arrays a and da are displayed with only two. The columns of the displayed arrays need not line up as they do above.

4. Write a script to read the data from the previous problem into three NumPy arrays with the variable names f, a, da for the frequency, amplitude, and amplitude error and then, in the same script, write the data out to a data file, including the header, with the data displayed in three columns, just as it's displayed in the problem above. It's ok if the header lines begin with the # comment character. Your data file should have the extension .txt.

5. Write a script to read the data from the previous problem into three NumPy arrays with the variable names f, a, da for the frequency, amplitude, and amplitude error and then, in the same script, write the data out to a csv data file, without the header, to a data file with the data displayed in three columns. Use a single format specifier and set it to "%0.16e". If you have access the

spreadsheet program (like MS Excel), try opening the file you have created with your Python script and verify that the arrays are displayed in three columns. Note that your csv file should have the extension .csv.

CHAPTER 5

Conditionals and Loops

> In this chapter, you learn how to **control the flow of a program**. In particular, you learn how to make a computer make decisions based on information it receives from you or based on different conditions it encounters as it processes data or information. You also learn how to make a computer do repetitive tasks.

Computer programs are useful for performing repetitive tasks. Without complaining, getting bored, or growing tired, they can repetitively perform the same calculations with minor, but important, variations over and over again. Humans share with computers none of these qualities. And so we humans employ computers to perform the massive repetitive tasks we would rather avoid. However, we need efficient ways of telling the computer to do these repetitive tasks; we don't want to have stop to tell the computer each time it finishes one iteration of a task to do the task again, but for a slightly different case. We want to tell it once, "Do this task 1000 times with slightly different conditions and report back to me when you are done." This is what **loops** were made for.

In the course of doing these repetitive tasks, computers often need to make decisions. In general, we don't want the computer to stop and ask us what it should do if a certain result is obtained from its calculations. We might prefer to say, "Look, if you get result A during your calculations, do this, otherwise, do this other thing." That is, we often want to tell the computer ahead of time what to do if it encounters different situations. This is what **conditionals** were made for.

Conditionals and loops control the flow of a program. They are essential to performing virtually any significant computational task. Python, like most computer languages, provides a variety of ways of implementing loops and conditionals.

5.1 Conditionals

Conditional statements allow a computer program to take different actions based on whether some condition, or set of conditions is true or false. In this way, the programmer can control the flow of a program.

5.1.1 `if`, `elif`, and `else` statements

The `if`, `elif`, and `else` statements are used to define conditionals in Python. We illustrate their use with a few examples.

if-elif-else example

Suppose we want to know if the solutions to the quadratic equation

$$ax^2 + bx + c = 0$$

are real, imaginary, or complex for a given set of coefficients a, b, and c. Of course, the answer to that question depends on the value of the discriminant $d = b^2 - 4ac$. The solutions are real if $d \geq 0$, imaginary if $b = 0$ and $d < 0$, and complex if $b \neq 0$ and $d < 0$. The program below implements the above logic in a Python program.

Code: chapter5/programs/if-elif-elseExample.py

```python
a = float(input("What is the coefficient a? "))
b = float(input("What is the coefficient b? "))
c = float(input("What is the coefficient c? "))
d = b*b - 4.*a*c
if d >= 0.0:
    print("Solutions are real")        # block 1
elif b == 0.0:
    print("Solutions are imaginary")   # block 2
else:
    print("Solutions are complex")     # block 3
print("Finished")
```

After getting the inputs from the user, the program evaluates the discriminant d. The expression `d >= 0.0` has a Boolean truth value of either `True` or `False` depending on whether or not $d \geq 0$. You can check this out in the interactive IPython shell by typing the following set of commands

```
In [2]: d = 5

In [3]: d >= 2
```

Conditionals and Loops

Out[3]: **True**

In [4]: d >= 7
Out[4]: **False**

Therefore, the `if` statement in line 5 is simply testing to see if the statement `d >= 0.0` is `True` or `False`. If the statement is `True`, Python executes the indented block of statements following the `if` statement. In this case, there is only one line in indented block. Once it executes this statement, Python skips past the `elif` and `else` blocks and executes the `print("Finished!")` statement.

If the `if` statement in line 5 is `False`, Python skips the indented block directly below the `if` statement and executes the `elif` statement. If the condition `b == 0.0` is `True`, it executes the indented block immediately below the `elif` statement and then skips the `else` statement and the indented block below it. It then executes the `print("Finished!")` statement.

Finally, if the `elif` statement is `False`, Python skips to the `else` statement and executes the block immediately below the `else` statement. Once finished with that indented block, it then executes the `print("Finished!")` statement.

As you can see, each time a `False` result is obtained in an `if` or `elif` statement, Python skips the indented code block associated with that

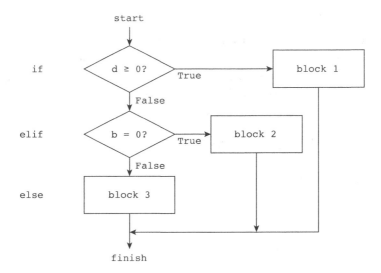

Figure 5.1 Flowchart for `if-elif-else` code.

statement and drops down to the next conditional statement, that is, the next `elif` or `else`. A flowchart of the if-elif-else code is shown in Fig. 5.1.

At the outset of this problem we stated that the solutions to the quadratic equation are imaginary only if $b = 0$ and $d < 0$. In the `elif b == 0.0` statement on line 7, however, we only check to see if $b = 0$. The reason that we don't have to check if $d < 0$ is that the `elif` statement is executed only if the condition `if d >= 0.0` on line 5 is `False`. Similarly, we don't have to check if $b = 0$ and $d < 0$ for the final `else` statement because this part of the `if`, `elif`, and `else` block will only be executed if the preceding `if` and `elif` statements are `False`. This illustrates a key feature of the `if`, `elif`, and `else` statements: these statements are executed sequentially until one of the `if` or `elif` statements is found to be `True`. Therefore, Python reaches an `elif` or `else` statement only if all the preceding `if` and `elif` statements are `False`.

The `if-elif-else` logical structure can accommodate as many `elif` blocks as desired. This allows you to set up logic with more than the three possible outcomes illustrated in the example above. When designing the logical structure you should keep in mind that once Python finds a true condition, it skips all subsequent `elif` and `else` statements in a given `if`, `elif`, and `else` block, irrespective of their truth values.

`if-else` example

You will often run into situations where you simply want the program to execute one of two possible blocks based on the outcome of an `if` statement. In this case, the `elif` block is omitted and you simply use an `if-else` structure. The following program testing whether an integer is even or odd provides a simple example.

Code: chapter5/programs/if-elseExample.py

```
1  a = int(input("Please input an integer: "))
2  if a % 2 == 0:
3      print("{0:0d} is an even number.".format(a))
4  else:
5      print("{0:0d} is an odd number.".format(a))
```

The flowchart in Fig. 5.2 shows the logical structure of an `if-else` structure.

Conditionals and Loops

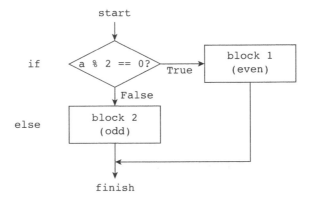

Figure 5.2   Flowchart for if-else code.

if example

The simplest logical structure you can make is a simple if statement, which executes a block of code if some condition is met but otherwise does nothing. The program below, which takes the absolute value of a number, provides a simple example of such a case.

Code: chapter5/programs/ifExample.py

```
1  a = eval(input("Please input a number: "))
2  if a < 0:
3      a = -a
4  print("The absolute value is {}".format(a))
```

When the block of code in an if or elif statement is only one line long, you can write it on the same line as the if or elif statement. For example, the above code can be written as follows.

Code: chapter5/programs/ifExampleAlt.py

```
1  a = eval(input("Please input a number: "))
2  if a < 0: a = -a
3  print("The absolute value is {}".format(a))
```

This works exactly as the preceding code. Note, however, that if the block of code associated with an if or elif statement is more than one line long, the entire block of code should be written as indented text below the if or elif statement.

The flowchart in Fig. 5.3 shows the logical structure of a simple if structure.

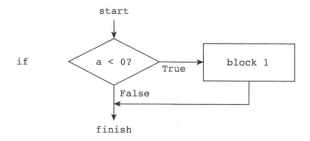

Figure 5.3   Flowchart for if code.

5.1.2   Logical operators

It is important to understand that "==" in Python is not the same as "=". The operator "=" is the assignment operator: d = 5 assigns the value of 5 to the variable d. On the other hand "==" is the *logical equals operator* and d == 5 is a *logical truth statement*. It tells Python to check to see if d is equal to 5 or not, and assigns a value of True or False to the statement d == 5 depending on whether or not d is equal to 5. Table 5.1 summarizes the various logical operators available in Python.

Operator	Function
<	less than
<=	less than or equal to
>	greater than
>=	greater than or equal to
==	equal
!=	not equal
and	both must be true
or	one or both must be true
not	reverses the truth value

Table 5.1   Logical operators in Python: comparisons (above double line), Boolean operations (below).

Table 5.1 lists three Boolean operators that we haven't encountered before: and, or, and not. These are useful for combining different logical conditions. For example, suppose you want to check if $a > 2$ and $b < 10$ simultaneously. To do so, you would write a > 2 and b < 10.

Conditionals and Loops

The code below illustrates the use of the logical operators `and`, `or`, and `not`.

```
In [6]: b = 10

In [7]: a != 5  # a is not equal to 5
Out[7]: False

In [8]: a>2 and b<20
Out[8]: True

In [9]: a>2 and b>10
Out[9]: False

In [10]: a>2 or b>10
Out[10]: True

In [11]: a>2
Out[11]: True

In [12]: not a>2
Out[12]: False
```

Logical statements like those above can be used in `if`, `elif`, and, as we shall see below, `while` statements, according to your needs.

5.2 Loops

In computer programming a *loop* is statement or block of statements that is executed repeatedly. Python has two kinds of loops, a `for` loop and a `while` loop. We first introduce the `for` loop and illustrate its use for a variety of tasks. We then introduce the `while` loop and, after a few illustrative examples, compare the two kinds of loops and discuss when to use one or the other.

5.2.1 `for` loops

The general form of a `for` loop in Python is

```
for <itervar> in <sequence>:
    <body>
```

where `<intervar>` is a variable, `<sequence>` is a sequence such as list or string or array, and `<body>` is a series of Python commands to be executed repeatedly for each element in the `<sequence>`. The `<body>`

is indented from the rest of the text, which defines the extent of the loop. Let's look at a few examples.

Code: chapter5/programs/doggyLoop.py

```
1  for dogname in ["Molly", "Max", "Buster", "Lucy"]:
2      print(dogname)
3      print(" Arf, arf!")
4  print("All done.")
```

Running this program, stored in file `doggyloop.py`, produces the following output.

```
In [1]: run doggyloop.py
Molly
 Arf, arf!
Max
 Arf, arf!
Buster
 Arf, arf!
Lucy
 Arf, arf!
All done.
```

The `for` loop works as follows: the *iteration variable* or *loop index* `dogname` is set equal to the first element in the list, `"Max"`, and then the two lines in the indented body are executed. Then `dogname` is set equal to second element in the list, `"Molly"`, and the two lines in the indented body are executed. The loop cycles through all the elements of the list, and then moves on to the code that follows the `for` loop and prints `All done`.

When indenting a block of code in a Python `for` loop, it is critical that every line be indented by the same amount. Using the **<tab>** key causes the Code Editor to indent 4 spaces. Any amount of indentation works, as long as it is the same for all lines in a `for` loop. While code editors designed to work with Python (including Spyder) translate the **<tab>** key to 4 spaces, not all text editors do. In those cases, 4 spaces are not equivalent to a **<tab>** character even if they appear the same on the display. Indenting some lines by 4 spaces and other lines by a **<tab>** character will produce an error. So beware!

Figure 5.4 shows the flowchart for a `for` loop. It starts with an implicit conditional asking if there are any more elements in the sequence. If there are, it sets the iteration variable equal to the next element in the sequence and then executes the body—the indented text—using that value of the iteration variable. It then returns to the

Conditionals and Loops

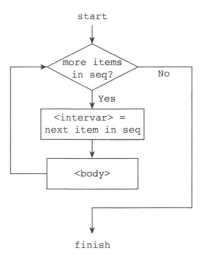

Figure 5.4   Flowchart for for-loop.

beginning to see if there are more elements in the sequence and continues the loop until there is none remaining.

Using an accumulator to calculate a sum

Let's look at another application of Python's for loop. Suppose you want to calculate the sum of all the odd numbers between 1 and 100. Before writing a computer program to do this, let's think about how you would do it by hand. You might start by adding 1+3=4. Then take the result 4 and add the next odd integer, 5, to get 4+5=9; then 9+7=16, then 16+9=25, and so forth. You are doing repeated additions, starting with 1+3, while keeping track of the running sum, until you reach the last number 99.

In developing an algorithm for having the computer sum the series of numbers, we are going to do the same thing: add the numbers one at a time while keeping track of the running sum, until we reach the last number. We will keep track of the running sum with the variable s, which is called the *accumulator*. Initially s = 0, since we haven't added any numbers yet. Then we add the first number, 1, to s and s becomes 1. Then we add the next number, 3, in our sequence of odd numbers to s and s becomes 4. We continue doing this over and over again using a for loop while the variable s accumulates the running sum until we reach the final number. The code below illustrates how to do this.

Code: chapter5/programs/oddSum100.py

```
1  s = 0
2  for i in range(1, 100, 2):
3      print(i, end=' ')
4      s = s+i
5  print('\n{}'.format(s))
```

The `range` function defines the sequence of odd numbers 1, 3, 5, ..., 97, 99. The `for` loop successively adds each number in the list to the running sum until it reaches the last element in the list and the sum is complete. Once the `for` loop finishes, the program exits the loop and prints final value of `s`, which is the sum of the odd numbers from 1 to 99, is printed out. Line 3 is not needed, of course, and is included only to verify that the odd numbers between 1 and 100 are being summed. The `end=' '` argument causes a space to be printed out between each value of `i` instead of the default new line character `\n`. Copy the above program and run it. You should get an answer of 2500.

As noted on page 39, the `range` produces an *iterable sequence*, which is a set of instructions that yields the next value in a sequence of integers each time it is accessed.

Iterating over sequences

Thus far we have seen that `for` loops can iterate over elements in a list, such as the names of dogs in our first loop example, or over a sequence of numbers produced by the `range` function. In fact, Python `for` loops are extremely versatile and can be used with any object that consists of a sequence of elements. Some of the ways it works might surprise you. For example, suppose we have the string

```
In [1]: a = 'There are places I remember all my life'
```

The string `a` is a sequence of characters and thus can be looped over, as illustrated here:

```
In [2]: for letter in a:
   ...:     print(letter)
   ...:
T
h
e
r
```

Conditionals and Loops

```
e
a
r
.
.
.
```

Suppose I wanted to print out every third letter of this string. One way to do it would be to set up a counter, as follows:

```
In [3]: i = 0
   ...:     for letter in a:
   ...:         if i % 3 == 0:
   ...:             print(letter, end=' ')
   ...:         i += 1
   ...:
T r a    a s    m b    l y i
```

While this approach works just fine, Python has a function called `enumerate` that does it for you. It works like this:

```
In [4]: for i, letter in enumerate(a):
   ...:     if i % 3 == 0:
   ...:         print(letter, end=' ')
   ...:
T r a    a s    m b    l y i
```

The `enumerate` function takes two inputs, a counter (`i` in this case) and a sequence (the string `a`). In general, any kind of sequence can be used in place of `a`, a list or a NumPy array, for example. Pretty slick. We will find plenty of opportunities to use the `enumerate` function.

range, enumerate, strings, lines in files, ...

5.2.2 `while` loops

The general form of a `while` loop in Python is

```
while <condition>:
    <body>
```

where `<condition>` is a statement that can be either `True` or `False` and `<body>` is a series of Python commands that is executed repeatedly until `<condition>` becomes false. This means that somewhere in `<body>`, the truth value of `<condition>` must be changed so that it becomes false after a finite number of iterations. Consider the following example.

Suppose you want to calculate all the Fibonacci numbers smaller than 1000. The Fibonacci numbers are determined by starting with the integers 0 and 1. The next number in the sequence is the sum of the previous two. So, starting with 0 and 1, the next Fibonacci number is $0 + 1 = 1$, giving the sequence $0, 1, 1$. Continuing this process, we obtain $0, 1, 1, 2, 3, 5, 8, \ldots$ where each element in the list is the sum of the previous two. Using a `for` loop to calculate the Fibonacci numbers is impractical because we do not know ahead of time how many Fibonacci numbers there are smaller than 1000. By contrast a `while` loop is perfect for calculating all the Fibonacci numbers because it keeps calculating Fibonacci numbers until it reaches the desired goal, in this case 1000. Here is the code using a `while` loop.

```
x, y = 0, 1
while x < 1000:
    print(x)
    x, y = y, x+y
```

We have used the multiple assignment feature of Python in this code. Recall that all the values on the left are assigned using the current values of `x` and `y` on the right.

Figure 5.5 shows the flowchart for the `while` loop. The loop starts with the evaluation of a condition. If the condition is `False`, the code in the body is skipped, the flow exits the loop, and then continues with the rest of the program. If the condition is `True`, the code in the body—the indented text—is executed. Once the body is finished, the flow returns to the condition and proceeds along the `True` or `False` branches depending on the truth value of the condition. Implicit in this loop is the idea that somewhere during the execution of the body of the while loop, the variable that is evaluated in the condition is changed in some way. Eventually that change will cause the condition to return a value of `False` so that the loop will end.

One danger of a `while` loop is that it is entirely possible to write a loop that never terminates—an *infinite loop*. For example, if we had written `while y > 0:`, in place of `while x < 1000:`, the loop would never end. If you execute code that has an infinite loop, you can often terminate the program from the keyboard by typing **ctrl-C** a couple of times. If that doesn't work, you may have to terminate and then restart Python.

For the kind of work we do in science and engineering, we generally find that the `for` loop is more useful than the `while` loop. Never-

Conditionals and Loops

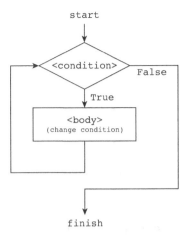

Figure 5.5 Flowchart for `while` loop.

theless, there are times when using a `while` loop is better suited to a task than is a `for` loop.

5.2.3 Loops and array operations

Loops are often used to sequentially modify the elements of an array. For example, suppose we want to square each element of the array `a = np.linspace(0, 32, 1e7)`. This is a hefty array with 10 million elements. Nevertheless, the following loop does the trick.

Code: chapter5/programs/slowLoops.py

```
1  import numpy as np
2  import time
3  a = np.linspace(0, 32, 10000000)   # 10 million
4  print(a)
5  startTime = time.process_time()
6  for i in range(len(a)):
7      a[i] = a[i]*a[i]
8  endTime = time.process_time()
9  print(a)
10 print('Run time = {} seconds'.format(endTime-startTime))
```

Running this on my computer returns the result in about 3.4 seconds—not bad for having performed 10 million multiplications. Notice that we have introduced the `time` module, which we use to measure how long (in seconds) it takes the computer to perform the 10 million multiplications.

Of course we could have performed the same calculation using the array multiplication we learned in Chapter 3. To do so, we replace the for loop in lines 6–7 above with a simple array multiplication in line 6 below. Here is the code.

Code: chapter5/programs/fastArray.py

```
import numpy as np
import time
a = np.linspace(0, 32, 10000000)   # 10 million
print(a)
startTime = time.process_time()
a = a*a
endTime = time.process_time()
print(a)
print('Run time = {} seconds'.format(endTime-startTime))
```

Running this on my computer returns the results in about 1/50 of a second, more than 100 times faster than we obtained using a loop. This illustrates an important point: **for loops are slow**. Array operations run much faster and are therefore to be preferred in any case where you have a choice. Sometimes finding an array operation that is equivalent to a loop can be difficult, especially for a novice. Nevertheless, doing so pays rich rewards in execution time. Moreover, the array notation is usually simpler and clearer, providing further reasons to prefer array operations over loops.

5.3 List Comprehensions

List comprehensions are a special feature of core Python for processing and constructing lists. We introduce them here because they use a looping process. They are used quite commonly in Python coding and they often provide elegant compact solutions to some common computing tasks.

Consider, for example the 3 ×3 matrix

```
In [1]: A = [[1, 2, 3],
   ...:      [4, 5, 6],
   ...:      [7, 8, 9]]
```

Suppose we want to construct a vector from the diagonal elements of this matrix. We could do so with a for loop with an accumulator as follows:

```
In [2]: diag = []
In [3]: for i in [0, 1, 2]:
   ...:     diag.append(A[i][i])
   ...:
```

```
In [4]: diag
Out[4]: [1, 5, 9]
```

Here we have used the `append()` list method to add elements to the list `diag` one at a time.

List comprehensions provide a simpler, cleaner, and faster way to build a list of the diagonal elements of A:

```
In [5]: diagLC = [A[i][i] for i in [0, 1, 2]]

In [6]: diagLC
Out[6]: [1, 5, 9]
```

A one-line list comprehension replaces a three-line accumulator plus loop code. Suppose we now want the square of this list:

```
In [7]: [y*y for y in diagLC]
Out[7]: [1, 25, 81]
```

Notice here how y serves as a dummy variable accessing the various elements of the list `diagLC`.

Extracting a row from a 2-dimensional array such as A is quite easy. For example the second row is obtained quite simply in the following fashion

```
In [8]: A[1]
Out[8]: [4, 5, 6]
```

Obtaining a column is not as simple, but a list comprehension makes it quite straightforward:

```
In [9]: c1 = [a[1] for a in A]
In [10]: c1
Out[10]: [2, 5, 8]
```

Another, slightly less elegant way to accomplish the same thing is

```
In [11]: [A[i][1] for i in range(3)]
Out[11]: [2, 5, 8]
```

Suppose you have a list of numbers and you want to extract all the elements of the list that are divisible by three. A slightly fancier list comprehension accomplishes the task quite simply and demonstrates a new feature:

```
In [12]: y = [-5, -3, 1, 7, 4, 23, 27, -9, 11, 41]
In [13]: [x for x in y if x%3==0]
Out[13]: [-3, 27, -9]
```

As we see in this example, a conditional statement can be added to a list comprehension. Here it serves as a filter to select out only those elements that are divisible by three.

5.4 Exercises

1. Write a program to calculate the factorial of a positive integer input by the user. Recall that the factorial function is given by $x! = x(x-1)(x-2)\ldots(2)(1)$ so that $1! = 1$, $2! = 2$, $3! = 6$, $4! = 24$, $5! = 120, \ldots$

    (a) Write the factorial function using a Python `while` loop.

    (b) Write the factorial function using a Python `for` loop.

    Check your programs to make sure they work for 1, 2, 3, 5, and beyond, but especially for the first 5 integers.

2. The following Python program finds the smallest non-trivial (not 1) prime factor of a positive integer.

    ```
    n = int(input("Input an integer > 1: "))
    i = 2
    while (n % i) != 0:
        i += 1
    print("The smallest factor of {0:d} is {1:d}".format(n, i))
    ```

    (a) Type this program into your computer and verify that it works as advertised. Then briefly explain how it works and why the `while` loop always terminates.

    (b) Modify the program so that it tells you if the integer input is a prime number or not. If it is not a prime number, write your program so that it prints out the smallest prime factor. Using your program verify that the following integers are prime numbers: 101, 8191, 94811, 947431.

3. Consider the matrix list x = [[1, 2, 3], [4, 5, 6], [7, 8, 9]]. Write a list comprehension to extract the last column of the matrix [3, 6, 9]. Write another list comprehension to create a vector of twice the square of the middle column [8, 50, 128].

4. Write a program that calculates the value of an investment after some number of years specified by the user if

Conditionals and Loops

(a) the principal is compounded annually

(b) the principle is compounded monthly

(c) the principle is compounded daily

Your program should ask the user for the initial investment (principal), the interest rate in percent, and the number of years the money will be invested (allow for fractional years). For an initial investment of $1000 at an interest rate of 6%, after 10 years I get $1790.85 when compounded annually, $1819.40 when compounded monthly, and $1822.03 when compounded daily, assuming 12 months in a year and 365.24 days in a year, where the monthly interest rate is the annual rate divided by 12 and the daily rate is the annual rate divided by 365 (don't worry about leap years).

5. Write a program that determines the day of the week for any given calendar date after January 1, 1900, which was a Monday. Your program will need to take into account leap years, which occur in every year that is divisible by 4, except for years that are divisible by 100 but are not divisible by 400. For example, 1900 was not a leap year, but 2000 was a leap year. Test that your program gives the answers tabulated below.

Date	Weekday
January 1, 1900	Monday
June 28, 1919	Saturday
January 30, 1928	Tuesday
December 5, 1933	Tuesday
February 29, 1948	Sunday
March 1, 1948	Monday
January 15, 1953	Thursday
November 22, 1963	Friday
June 23, 1993	Wednesday
August 28, 2005	Sunday
May 16, 2111	Saturday

CHAPTER 6

Plotting

> *We introduce **plotting** using the **matplotlib** package. You will learn how to make simple 2D x-y plots and fancy plots suitable for publication, complete with legends, annotations, logarithmic axes, subplots, and insets. You will learn how to produce **Greek letters and mathematical symbols** using a powerful markup language called LaTeX. You will also learn how to make various kinds of contour plots and vector field plots. We provide an introduction to 3D plotting with matplotlib, although its 3D capabilities are somewhat limited. We begin with a simple but limited interface, known as **PyPlot**, but then move to the more powerful object-oriented interface. We also provide an overview of the object-oriented structure of matplotlib, including the important but sometimes confusing topic of **backends**, which is helpful in reading online documentation and for understanding interactive plotting, which we do not cover in this book.*

The graphical representation of data—plotting—is one of the most important tools for evaluating and understanding scientific data and theoretical predictions. Plotting is not a part of core Python, however, but is provided through one of several possible library modules. The most highly developed and widely used plotting package for Python is matplotlib (http://matplotlib.sourceforge.net/). It is a powerful and flexible program that has become the *de facto* standard for 2D plotting with Python.

Because matplotlib is an external library—in fact it's a collection of libraries—it must be imported into any routine that uses it. matplotlib makes extensive use of NumPy so the two should be imported together. Therefore, for any program that is to produce 2D plots, you should include the lines

```
import numpy as np
import matplotlib.pyplot as plt
```

There are other matplotlib sub-libraries, but the `pyplot` library provides nearly everything that you need for 2D plotting. The standard prefix for it is `plt`. On some installations, matplotlib is automatically loaded with the IPython shell so you do not need to use `import matplotlib.pyplot` nor do you need to use the `plt` prefix when working in the IPython shell.[1]

One final word before we get started: We only scratch the surface of what is possible using matplotlib and as you become familiar with it, you will surely want to do more than this manual describes. In that case, you need to go the web to get more information. A good place to start is http://matplotlib.org/index.html. Another interesting web page is http://matplotlib.org/gallery.html.

6.1 An Interactive Session with PyPlot

We begin with an interactive plotting session that illustrates some very basic features of matplotlib. After importing NumPy and matplotlib, type in the `plot` command shown below and press the return key. Take care to follow the exact syntax.

```
In [1]: import numpy as np

In [2]: import matplotlib.pyplot as plt

In [3]: plt.plot([1, 2, 3, 2, 3, 4, 3, 4, 5])
Out[3]: [<matplotlib.lines.Line2D at 0x94e1310>]

In [4]:  plt.show()
```

A window should appear with a plot that looks something like the interactive plot window shown in Fig. 6.1.[2] By default, the `plot` function draws a line between the data points that were entered. You can save this figure to an image file by clicking on the floppy disk (Save the figure) icon 🖫 at the top of the plot window. You can also zoom 🔍, pan ✥, scroll through the plot, and return to the original

---

[1] This can depend on the settings in your particular installation. You may want to set the Preferences for the IPython console to automatically load PyLab, which loads the NumPy and matplotlib modules. See §A.1.1 of Appendix A.

[2] Here we have assumed that matplotlib's *interactive mode* is turned off. If it's turned on, then you don't need the `plt.show()` function when plotting from the command line of the IPython shell. Type `plt.ion()` at the IPython prompt to turn on interactive mode. Type `plt.ioff()` to turn it off. Whether or not you work with interactive mode turned on in IPython is largely a matter of taste.

Plotting

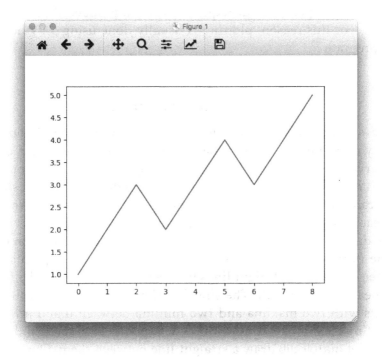

Figure 6.1   Interactive plot window.

view 🏠 using other icons in the plot window. Experimenting with them reveals their functions. See page 118 for information about the configure subplots icon. When you are finished, be sure to close the plot window, which will return control to the IPython console.

Let's take a closer look at the `plot` function. It is used to plot *x*-*y* data sets and is written like this

```
plot(x, y)
```

where `x` and `y` are arrays (or lists) that have the same size. If the `x` array is omitted, that is, if there is only a single array, as in our example above, the `plot` function uses 0, 1, ..., N-1 for the `x` array, where N is the size of the `y` array. Thus, the `plot` function provides a quick graphical way of examining a data set.

More typically, you supply both an *x* and a *y* data set to plot. Taking things a bit further, you may also want to plot several data sets on the same graph, use symbols as well as lines, label the axes, create a title and a legend, and control the color of symbols and lines.

All of this is possible but requires calling a number of plotting functions. For this reason, plotting is usually done using a Python script or program.

6.2 Basic Plotting

The quickest way to learn how to plot using the matplotlib library is by example. For our first task, let's plot the sine function over the interval from 0 to 4π. The main plotting function `plot` in matplotlib does not plot functions *per se*, it plots (x,y) data sets. As we shall see, we can instruct the function `plot` either to just draw points—or dots—at each data point, or we can instruct it to draw straight lines between the data points. To create the illusion of the smooth function that the sine function is, we need to create enough (x,y) data points so that when `plot` draws straight lines between the data points, the function appears to be smooth. The sine function undergoes two full oscillations with two maxima and two minima between 0 and 4π. So let's start by creating an array with 33 data points between 0 and 4π, and then let matplotlib draw a straight line between them. Our code consists of four parts:

- Import the NumPy and matplotlib modules (lines 1-2 below).

- Create the (x,y) data arrays (lines 3-4 below).

- Have `plot` draw straight lines between the (x,y) data points (line 5 below).

- Display the plot in a figure window using the `show` function (line 6 below).

Here is our code, which consists of only 6 lines:

Code: chapter6/programs/sineFunctionPlot33.py

```
import numpy as np
import matplotlib.pyplot as plt
x = np.linspace(0, 4.*np.pi, 33)
y = np.sin(x)
plt.plot(x, y)
plt.show()
```

Only 6 lines suffice to create the plot, which is shown on the left side of Fig. 6.2. It consists of the sine function plotted over the interval

Plotting 103

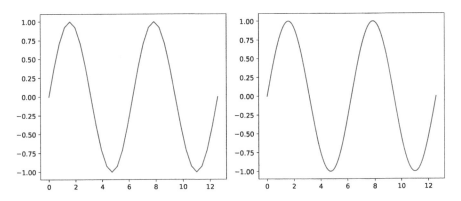

Figure 6.2 Sine function: LEFT, 33 data points; RIGHT, 129 data points.

from 0 to 4π, as advertised, as well as axes annotated with nice whole numbers over the appropriate interval. It's a pretty nice plot made with very little code.

One problem, however, is that while the plot oscillates like a sine wave, it is not smooth (look at the peaks). This is because we did not create the (x, y) arrays with enough data points. To correct this, we need more data points. The plot on the right side of Fig. 6.2 was created using the same program shown above but with 129 (x, y) data points instead of 33. Try it out yourself by copying the above program and replacing 33 in line 3 with 129 (a few more or less is ok) so that the function `linspace` creates an array with 129 data points instead of 33.

The above script illustrates how plots can be made with very little code using the matplotlib module. In making this plot, matplotlib has made a number of choices, such as the size of the figure, the color of the line, even the fact that by default a line is drawn between successive data points in the (x, y) arrays. All of these choices can be changed by explicitly instructing matplotlib to do so. This involves including more arguments in the function calls we have used and using new functions that control other properties of the plot. The next example illustrates a few of the simpler embellishments that are possible.

In Fig. 6.3, we plot two (x, y) data sets: a smooth line curve and some data represented by red circles. In this plot, we label the x and y axes, create a legend, and draw lines to indicate where x and y are zero. The code that creates this plot is shown in Fig. 6.3.

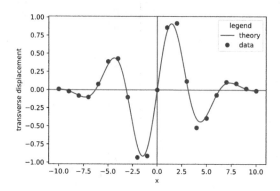

Figure 6.3   Wavy pulse.

Code: chapter6/programs/wavyPulse.py

```
1  import numpy as np
2  import matplotlib.pyplot as plt
3
4  # read data from file
5  xdata, ydata = np.loadtxt('wavyPulseData.txt', unpack=True)
6
7  # create x and y arrays for theory
8  x = np.linspace(-10., 10., 200)
9  y = np.sin(x) * np.exp(-(x/5.0)**2)
10
11 # create plot
12 plt.figure(1, figsize=(6, 4))
13 plt.plot(x, y, 'g-', label='theory')        # 'g-' green line
14 plt.plot(xdata, ydata, 'bo', label="data")  # 'bo' blue circles
15 plt.xlabel('x')
16 plt.ylabel('transverse displacement')
17 plt.legend(loc='upper right', title='legend')
18 plt.axhline(color='gray', zorder=-1)
19 plt.axvline(color='gray', zorder=-1)
20
21 # save plot to file
22 plt.savefig('figures/WavyPulse.pdf')
23
24 # display plot on screen
25 plt.show()
```

If you have read the first five chapters, the code in lines 1–9 in the above script should be familiar to you. First, the script loads the NumPy and matplotlib modules, then reads data from a data file into two arrays, xdata and ydata, and then creates two more arrays, x and

y. The first pair or arrays, xdata and ydata, contain the *x-y* data that are plotted as blue circles in Fig. 6.3; the arrays created in lines 8 and 9 contain the *x-y* data that are plotted as a green line.

The functions that create the plot begin on line 12. Let's go through them one by one and see what they do. You will notice that *keyword arguments* (kwargs) are used in several cases.

figure() creates a blank figure window. If it has no arguments, it creates a window that is 8 inches wide and 6 inches high by default, although the size that appears on your computer depends on your screen's resolution. For most computers, it will be much smaller. You can create a window whose size differs from the default using the optional keyword argument figsize, as we have done here. If you use figsize, set it equal to a 2-element tuple where the elements, expressed in inches, are the width and height, respectively, of the plot. Multiple calls to figure() opens multiple windows: figure(1) opens up one window for plotting, figure(2) another, and figure(3) yet another.

plot(x, y, *optional arguments*) graphs the *x-y* data in the arrays x and y. The third argument is a format string that specifies the color and the type of line or symbol that is used to plot the data. The string 'bo' specifies a blue (b) circle (o). The string 'g-' specifies a green (g) solid line (-). The keyword argument label is set equal to a string that labels the data if the legend function is called subsequently.

xlabel(*string*) takes a string argument that specifies the label for the graph's *x*-axis.

ylabel(*string*) takes a string argument that specifies the label for the graph's *y*-axis.

legend() makes a legend for the data plotted. Each *x-y* data set is labeled using the string that was supplied by the label keyword in the plot function that graphed the data set. The loc keyword argument specifies the location of the legend. The title keyword can be used to give the legend a title.

axhline() draws a horizontal line across the width of the plot at y=0. Writing axhline(y=a) draws a horizontal line at y=a, where y=a can be any numerical value. The optional keyword argument color is a string that specifies the color of the line. The default color is black. The optional keyword argument `zorder` is an integer that specifies which plotting elements are in front of or behind others. By default, new plotting elements appear *on top of* previously plotted elements and have a value of `zorder=0`. By specifying `zorder=-1`, the horizontal line is plotted *behind* all existing plot elements that have not be assigned an explicit `zorder` less than −1. The keyword `zorder` can also be used as an argument for the `plot` function to specify the order of lines and symbols. Normally, for example, symbols are placed on top of lines that pass through them.

axvline() draws a vertical line from the top to the bottom of the plot at x=0. See `axhline()` for an explanation of the arguments.

savefig(*string*) saves the figure to a file with a name specified by the string argument. The string argument can also contain path information if you want to save the file someplace other than the default directory. Here we save the figure to a subdirectory named `figures` of the default directory. The extension of the filename determines the format of the figure file. The following formats are supported: png, pdf, ps, eps, and svg.

show() displays the plot on the computer screen. No screen output is produced before this function is called.

To plot the solid blue line, the code uses the `'b-'` format specifier in the `plot` function call. It is important to understand that matplotlib draws *straight lines* between data points. Therefore, the curve will appear smooth only if the data in the NumPy arrays are sufficiently dense. If the space between data points is too large, the straight lines the `plot` function draws between data points will be visible. For plotting a typical function, something on the order of 100–200 data points usually produces a smooth curve, depending on just how curvy the function is. On the other hand, only two points are required to draw a smooth straight line.

Detailed information about the matplotlib plotting functions are available online. The main matplotlib site is http://matplotlib.org/.

6.2.1 Specifying line and symbol types and colors

In the above example, we illustrated how to draw one line type (solid), one symbol type (circle), and two colors (blue and red). There are

Plotting

many more possibilities: some of which are specified in Table 6.1 and Fig. 6.4. The way it works is to specify a string consisting of up to three format specifiers: one for the symbol, one for line type, and another for color. It does not matter in which order the format specifiers are listed in the string. Examples are given following the two tables. Try them out to make sure you understand how these plotting format specifiers work.

Table 6.1 shows the characters used to specify the line or symbol type that is used. If a line type is chosen, the lines are drawn between the data points. If a marker type is chosen, the marker is plotted at each data point.

character	description	character	description
-	solid line style	3	tri_left marker
--	dashed line style	4	tri_right marker
-.	dash-dot line style	s	square marker
:	dotted line style	p	pentagon marker
.	point marker	*	star marker
,	pixel marker	h	hexagon1 marker
o	circle marker	H	hexagon2 marker
v	triangle_down marker	+	plus marker
^	triangle_up marker	x	x marker
<	triangle_left marker	D	diamond marker
>	triangle_right marker	d	thin_diamond marker
1	tri_down marker	|	vline marker
2	tri_up marker	_	hline marker

Table 6.1 Line and symbol type designations for plotting.

Color is specified using the codes in Fig. 6.4: single letters for primary colors and codes C0, C2, ..., C9 for a standard matplotlib color palette of ten colors designed to be pleasing to the eye.

Here are some examples of how these format specifiers can be used:

```
plot(x, y, 'ro')     # red circles
plot(x, y, 'ks-')    # black squares connected by black lines
plot(x, y, 'g^')     # green triangles pointing up
plot(x, y, 'k-')     # black line
plot(x, y, 'C1s')    # orange(ish) squares
```

These format specifiers give rudimentary control of the plotting

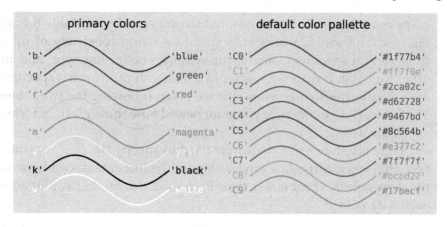

Figure 6.4 Some matplotlib colors. The one- or two-letter strings to the left of each curve, or the longer strings to the right of each curve, can be used to specify a designated color. However, the shades of cyan, magenta, and yellow for the one-letter codes are different from the full-word codes (shown).

symbols and lines. matplotlib provides much more precise control of the plotting symbol size, line types, and colors using optional keyword arguments instead of the plotting format strings introduced above. For example, the following command creates a plot of large yellow diamond symbols with orange edges connected by a green dashed line:

```
plot(x, y, color='green', linestyle='dashed', marker='D',
    markerfacecolor='yellow', markersize=7,
    markeredgecolor='C1')
```

Try it out! Another useful keyword is `fillstyle`, with self-explanatory keywords `full` (the default), `left`, `right`, `bottom`, `top`, `none`. The online matplotlib documentation provides all the plotting format keyword arguments and their possible values.

6.2.2 Error bars

When plotting experimental data it is customary to include error bars that indicate graphically the degree of uncertainty that exists in the measurement of each data point. The matplotlib function `errorbar` plots data with error bars attached. It can be used in a way that either replaces or augments the `plot` function. Both vertical and horizontal error bars can be displayed. Figure 6.5 illustrates the use of error bars.

Plotting

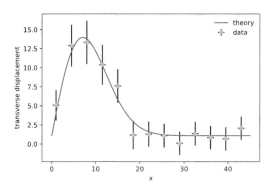

Figure 6.5 Data with error bars.

When error bars are desired, you typically replace the `plot` function with the `errorbar` function. The first two arguments of the `errorbar` function are the x and y arrays to be plotted, just as for the `plot` function. The keyword `fmt` *must be used* to specify the format of the points to be plotted; the format specifiers are the same as for `plot`. The keywords `xerr` and `yerr` are used to specify the x and y error bars. Setting one or both of them to a constant specifies one size for all the error bars. Alternatively, setting one or both of them equal to an array that has the same length as the x and y arrays allows you to give each data point an error bar with a different value. If you only want y error bars, then you should only specify the `yerr` keyword and omit the `xerr` keyword. The color of the error bars is set with the keyword `ecolor`.

The code below illustrates how to make error bars and was used to make the plot in Fig. 6.5. Lines 15 and 16 contain the call to the `errorbar` function. The x error bars are all set to a constant value of 0.75, meaning that the error bars extend 0.75 to the left and 0.75 to the right of each data point. The y error bars are set equal to an array, which was read in from the data file containing the data to be plotted, so each data point has a different y error bar. By the way, leaving out the `xerr` keyword argument in the `errorbar` function call below would mean that only the y error bars would be plotted.

Code: chapter6/programs/errorBarPlot.py

```
1  import numpy as np
2  import matplotlib.pyplot as plt
```

```
3
4   # read data from file
5   xdata, ydata, yerror = np.loadtxt('expDecayData.txt',
6                                    unpack=True)
7
8   # create theoretical fitting curve
9   x = np.linspace(0, 45, 128)
10  y = 1.1 + 3.0*x*np.exp(-(x/10.0)**2)
11
12  # create plot
13  plt.figure(1, figsize=(6, 4))
14  plt.plot(x, y, '-C0', label="theory")
15  plt.errorbar(xdata, ydata, fmt='oC1', label="data",
16               xerr=0.75, yerr=yerror, ecolor='black')
17  plt.xlabel('x')
18  plt.ylabel('transverse displacement')
19  plt.legend(loc='upper right')
20
21  # save plot to file
22  plt.savefig('figures/ExpDecay.pdf')
23
24  # display plot on screen
25  plt.show()
```

We have more to say about the errorbar function in the sections on logarithmic plots. But the brief introduction given here should suffice for making most plots not involving logarithmic axes.

6.2.3 Setting plotting limits and excluding data

It turns out that you often want to restrict the range of numerical values over which you plot data or functions. In these cases you may need to manually specify the plotting window or, alternatively, you may wish to exclude data points that are outside some set of limits. Here we demonstrate methods for doing this.

Setting plotting limits

Suppose you want to plot the tangent function over the interval from 0 to 10. The following script offers an straightforward first attempt.

Code: chapter6/programs/tanPlot0.py

```
1  import numpy as np
2  import matplotlib.pyplot as plt
3
4  theta = np.arange(0.01, 10., 0.04)
```

Plotting

```
5   ytan = np.tan(theta)
6
7   plt.figure(figsize=(8.5, 4.2))
8   plt.plot(theta, ytan)
9   plt.savefig('figures/tanPlot0.pdf')
10  plt.show()
```

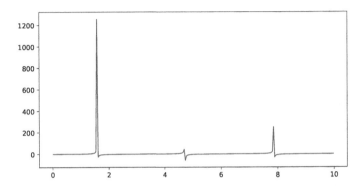

Figure 6.6 Trial tangent plot.

The resulting plot, shown in Fig. 6.6, doesn't quite look like what you might have expected for tan θ vs. θ. The problem is that tan θ diverges at $\theta = \pi/2, 3\pi/2, 5\pi/2, ...$, which leads to large spikes in the plots as values in the theta array come near those values. Of course, we don't want the plot to extend all the way out to $\pm\infty$ in the y direction, nor can it. Instead, we would like the plot to extend far enough that we get the idea of what is going on as $y \to \pm\infty$, but we would still like to see the behavior of the graph near $y = 0$. We can restrict the range of ytan values that are plotted using the matplotlib function ylim, as we demonstrate in the script below.

Code: chapter6/programs/tanPlot1.py

```
1   import numpy as np
2   import matplotlib.pyplot as plt
3
4   theta = np.arange(0.01, 10., 0.04)
5   ytan = np.tan(theta)
6
7   plt.figure(figsize=(8.5, 4.2))
8   plt.plot(theta, ytan)
9   plt.ylim(-8, 8)    # restricts range of y axis from -8 to +8
10  plt.axhline(color="gray", zorder=-1)
```

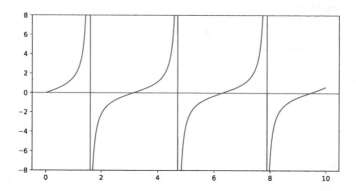

Figure 6.7 Tangent function (with spurious vertical lines).

```
11  plt.savefig('figures/tanPlot1.pdf')
12  plt.show()
```

Figure 6.7 shows the plot produced by this script, which now looks much more like the familiar $\tan\theta$ function we know. We have also included a call to the `axline` function to create an x axis.

Recall that for $\theta = \pi/2$, $\tan\theta^- \to +\infty$ and $\tan\theta^+ \to -\infty$; in fact, $\tan\theta$ diverges to $\pm\infty$ at every odd half integral value of θ. Therefore, the vertical blue lines at $\theta = \pi/2, 3\pi/2, 5\pi/2$ should not appear in a proper plot of $\tan\theta$ vs. θ. However, they do appear because the `plot` function simply draws lines between the data points in the x-y arrays provided the `plot` function's arguments. Thus, plot draws a line between the very large positive and negative ytan values corresponding to the theta values on either side of $\pi/2$ where $\tan\theta$ diverges to $\pm\infty$. It would be nice to exclude that line.

Masked arrays

We can exclude the data points near $\theta = \pi/2$, $3\pi/2$, and $5\pi/2$ in the above plot, and thus avoid drawing the nearly vertical lines at those points, using NumPy's *masked array* feature. The code below shows how this is done and produces Fig. 6.9. The masked array feature is implemented in line 6 with a call to NumPy's `masked_where` function in the sub-module `ma` (masked array). It is called by writing `np.ma.masked_where`. The `masked_where` function works as follows. The first argument sets the condition for masking elements of the array; the array is specified by the second argument. In this case, the

Plotting

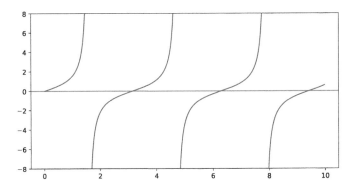

Figure 6.8 Tangent function (without spurious vertical lines).

function says to mask all elements of the array ytan (the second argument) where the absolute value of ytan is greater than 20. The result is set equal to ytanM. When ytanM is plotted, matplotlib's plot function omits all masked points from the plot. You can think of it as the plot function lifting the pen that is drawing the line in the plot when it comes to the masked points in the array ytanM.

Code: chapter6/programs/tanPlotMasked.py

```
1  import numpy as np
2  import matplotlib.pyplot as plt
3
4  theta = np.arange(0.01, 10., 0.04)
5  ytan = np.tan(theta)
6  ytanM = np.ma.masked_where(np.abs(ytan) > 20., ytan)
7
8  plt.figure(figsize=(8.5, 4.2))
9  plt.plot(theta, ytanM)
10 plt.ylim(-8, 8)    # restricts y-axis range from -8 to +8
11 plt.axhline(color="gray", zorder=-1)
12 plt.savefig('figures/tanPlotMasked.pdf')
13 plt.show()
```

6.2.4 Subplots

Often you want to create two or more graphs and place them next to one another, generally because they are related to each other in some way. Figure 6.9 shows an example of such a plot. In the top graph, $\tan\theta$ and $\sqrt{(8/\theta)^2 - 1}$ vs. θ are plotted. The two curves cross each

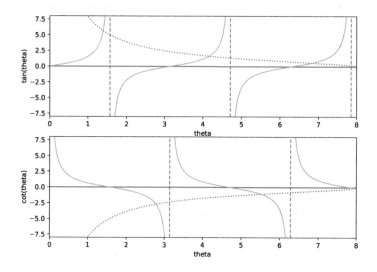

Figure 6.9 Plotting window with two subplots.

other at the points where $\tan\theta = \sqrt{(8/\theta)^2 - 1}$. In the bottom $\cot\theta$ and $-\sqrt{(8/\theta)^2 - 1}$ vs θ are plotted. These two curves cross each other at the points where $\cot\theta = -\sqrt{(8/\theta)^2 - 1}$.

The code that produces this plot is provided below.

Code: chapter6/programs/subplotDemo.py

```
import numpy as np
import matplotlib.pyplot as plt

theta = np.arange(0.01, 8., 0.04)
y = np.sqrt((8./theta)**2-1.)
ytan = np.tan(theta)
ytan = np.ma.masked_where(np.abs(ytan) > 20., ytan)
ycot = 1./np.tan(theta)
ycot = np.ma.masked_where(np.abs(ycot) > 20., ycot)

plt.figure(figsize=(8.5, 6))

plt.subplot(2, 1, 1)
plt.plot(theta, y, linestyle=':')
plt.plot(theta, ytan)
plt.xlim(0, 8)
plt.ylim(-8, 8)
plt.axhline(color="gray", zorder=-1)
```

```
19  plt.axvline(x=np.pi/2., color="gray", linestyle='--',
20              zorder=-1)
21  plt.axvline(x=3.*np.pi/2., color="gray", linestyle='--',
22              zorder=-1)
23  plt.axvline(x=5.*np.pi/2., color="gray", linestyle='--',
24              zorder=-1)
25  plt.xlabel("theta")
26  plt.ylabel("tan(theta)")
27
28  plt.subplot(2, 1, 2)
29  plt.plot(theta, -y, linestyle=':')
30  plt.plot(theta, ycot)
31  plt.xlim(0, 8)
32  plt.ylim(-8, 8)
33  plt.axhline(color="gray", zorder=-1)
34  plt.axvline(x=np.pi, color="gray", linestyle='--',
35              zorder=-1)
36  plt.axvline(x=2.*np.pi, color="gray", linestyle='--',
37              zorder=-1)
38  plt.xlabel("theta")
39  plt.ylabel("cot(theta)")
40
41  plt.savefig('figures/subplotDemo.pdf')
42  plt.show()
```

The function subplot, called on lines 13 and 28, creates the two subplots in the above figure. subplot has three arguments. The first specifies the number of rows into which the figure space is to be divided: in line 13, it's 2. The second specifies the number of columns into which the figure space is to be divided; in line 13, it's 1. The third argument specifies which rectangle will contain the plot specified by the following function calls. Line 13 specifies that the plotting commands that follow will act on the first box. Line 28 specifies that the plotting commands that follow will be act on the second box. As a convenience, the commas separating the three arguments in the subplot routine can be omitted, provided they are all single-digit arguments (less than or equal to 9). For example, lines 13 and 28 can be written as

```
plt.subplot(211)
   .
   .
plt.subplot(212)
```

Finally, we have also labeled the axes and included dashed vertical lines at the values of θ where $\tan\theta$ and $\cot\theta$ diverge.

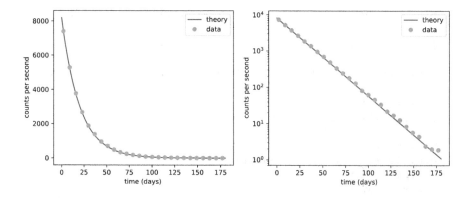

Figure 6.10 Semi-log plotting.

6.3 Logarithmic Plots

Data sets can span many orders of magnitude from fractional quantities much smaller than unity to values much larger than unity. In such cases it is often useful to plot the data on logarithmic axes.

6.3.1 Semi-log plots

For data sets that vary exponentially in the independent variable, it is often useful to use one or more logarithmic axes. Radioactive decay of unstable nuclei, for example, exhibits an exponential decrease in the number of particles emitted from the nuclei as a function of time. In Fig. 6.10, for example, we show the decay of the radioactive isotope Phosphorus-32 over a period of 6 months, where the radioactivity is measured once each week. Starting at a decay rate of nearly 10^4 electrons (counts) per second, the decay rate diminishes to only about 1 count per second after about 6 months or 180 days. If we plot counts per second as a function of time on a normal plot, as we have done in the left panel of Fig. 6.10, then the count rate is indistinguishable from zero after about 100 days. On the other hand, if we use a logarithmic axis for the count rate, as we have done in the right panel of Fig. 6.10, then we can follow the count rate well past 100 days and can readily distinguish it from zero. Moreover, if the data vary exponentially in time, then the data will fall along a straight line, as they do for the case of radioactive decay.

matplotlib provides two functions for making semi-logarithmic

Plotting

plots, `semilogx` and `semilogy`, for creating plots with logarithmic x and y axes, with linear y and x axes, respectively. We illustrate their use in the program below, which made the above plots.

Code: chapter6/programs/semilogDemo.py

```python
import numpy as np
import matplotlib.pyplot as plt

# read data from file
time, counts, unc = np.loadtxt('semilogDemo.txt',
                               unpack=True)

# create theoretical fitting curve
t_half = 14.              # P-32 half life = 14 days
tau = t_half/np.log(2)    # exponential tau
N0 = 8200.                # Initial count rate (per sec)
t = np.linspace(0, 180, 128)
N = N0 * np.exp(-t/tau)

# create plot
plt.figure(1, figsize=(9.5, 4))

plt.subplot(1, 2, 1)
plt.plot(t, N, color='C0', label="theory")
plt.plot(time, counts, 'oC1', label="data")
plt.xlabel('time (days)')
plt.ylabel('counts per second')
plt.legend(loc='upper right')

plt.subplot(1, 2, 2)
plt.semilogy(t, N, color='C0', label="theory")
plt.semilogy(time, counts, 'oC1', label="data")
plt.xlabel('time (days)')
plt.ylabel('counts per second')
plt.legend(loc='upper right')

plt.tight_layout()

# save plot to file
plt.savefig('figures/semilogDemo.pdf')

# display plot on screen
plt.show()
```

The `semilogx` and `semilogy` functions work the same way as the `plot` function. You just use one or the other depending on which axis you want to be logarithmic.

Adjusting spacing around subplots

You may have noticed the `tight_layout()` function, called without arguments on line 32 of the program. This is a convenience function that adjusts the space around the subplots to make room for the axes labels. If it is not called in this example, the y-axis label of the right plot runs into the left plot. The `tight_layout()` function can also be useful in graphics windows with only one plot sometimes.

If you want more control over how much space is allocated around subplots, use the function

```
plt.subplots_adjust(left=None, bottom=None, right=None,
                    top=None, wspace=None, hspace=None)
```

The keyword arguments `wspace` and `hspace` control the width and height of the space between plots, while the other arguments control the space to the left, bottom, right, and top. You can see and adjust the parameters of the `subplots_adjust()` routine by clicking on the configure subplots icon ≑ in the figure window. Once you have adjusted the parameters to obtain the desired effect, you can then use them in your script.

6.3.2 Log-log plots

matplotlib can also make log-log or double-logarithmic plots using the function `loglog`. It is useful when both the x and y data span many orders of magnitude. Data that are described by a power law $y = Ax^b$, where A and b are constants, appear as straight lines when plotted on a log-log plot. Again, the `loglog` function works just like the `plot` function but with logarithmic axes.

In the next section, we describe a more advanced syntax for creating plots, and with it an alternative syntax for making logarithmic axes. For a description, see page 122.

6.4 More Advanced Graphical Output

The plotting methods introduced in the previous sections are adequate for basic plotting but are recommended only for the simplest graphical output. Here, we introduce a more advanced syntax that harnesses the full power of matplotlib. It gives the user more options and greater control.

An efficient way to learn this new syntax is simply to look at an

Plotting

example. Figure 6.11, which shows multiple plots laid out in the same window, is produced by the following code:

Code: chapter6/programs/multiplePlots1window.py

```python
import numpy as np
import matplotlib.pyplot as plt


# Define the sinc function, with output for x=0
# defined as a special case to avoid division by zero
def s(x):
    a = np.where(x == 0., 1., np.sin(x)/x)
    return a


x = np.arange(0., 10., 0.1)
y = np.exp(x)

t = np.linspace(-15., 15., 150)
z = s(t)

# create a figure window
fig = plt.figure(figsize=(9, 7))

# subplot: linear plot of exponential
ax1 = fig.add_subplot(2, 2, 1)
ax1.plot(x, y, 'C0')
ax1.set_xlabel('time (ms)')
ax1.set_ylabel('distance (mm)')
ax1.set_title('exponential')

# subplot: semi-log plot of exponential
ax2 = fig.add_subplot(2, 2, 2)
ax2.plot(x, y, 'C2')
ax2.set_yscale('log')
# ax2.semilogy(x, y, 'C2')   # same as 2 previous lines
ax2.set_xlabel('time (ms)')
ax2.set_ylabel('distance (mm)')
ax2.set_title('exponential')

# subplot: wide subplot of sinc function
ax3 = fig.add_subplot(2, 1, 2)
ax3.plot(t, z, 'C3')
ax3.axhline(color='gray')
ax3.axvline(color='gray')
ax3.set_xlabel('angle (deg)')
ax3.set_ylabel('electric field')
ax3.set_title('sinc function')
```

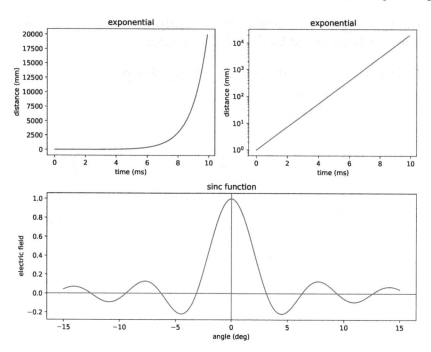

Figure 6.11 Mulitple plots in the same window.

```
45
46  # fig.tight_layout() adjusts white space to
47  # avoid collisions between subplots
48  fig.tight_layout()
49  fig.savefig("figures/multiplePlots1window.pdf")
50  fig.show()
```

After defining several arrays for plotting, the above program opens a figure window in line 19 with the statement

```
fig = plt.figure(figsize=(9, 7))
```

The matplotlib statement above creates a **Figure** object, assigns it the name fig, and opens a blank figure window. Thus, just as we give lists, arrays, and numbers variable names (*e.g.*, a = [1, 2, 5, 7], dd = np.array([2.3, 5.1, 3.9]), or st = 4.3), we can give a figure object and the window it creates a name: here it is fig. In fact we can use the figure function to open up multiple figure objects with different figure windows. The statements

```
fig1 = plt.figure()
fig2 = plt.figure()
```

Plotting

open up two separate windows, one named fig1 and the other fig2. We can then use the names fig1 and fig2 to plot things in either window. The figure function need not take any arguments if you are satisfied with the default settings such as the figure size and the background color. On the other hand, by supplying one or more keyword arguments, you can customize the figure size, the background color, and a few other properties. For example, in the program listing (line 25), the keyword argument figsize sets the width and height of the figure window. The default size is (8, 6); in our program we set it to (9, 8), which is a bit wider and higher than the default size. In the example above, we also choose to open only a single window, hence the single figure call.

The fig.add_subplot(2, 2, 1) in line 22 is a matplotlib function that divides the figure window into 2 rows (the first argument) and 2 columns (the second argument). The third argument creates a subplot in the first of the 4 subregions (*i.e.*, of the 2 rows × 2 columns) created by the fig.add_subplot(2, 2, 1) call. To see how this works, type the following code into a Python module and run it:

```
import numpy as np
import matplotlib.pyplot as plt

fig = plt.figure(figsize=(9, 8))
ax1 = fig.add_subplot(2,2,1)

fig.show()
```

You should get a figure window with axes drawn in the upper left quadrant. The fig. prefix used with the add_subplot(2, 2, 1) function directs Python to draw these axes in the figure window named fig. If we had opened two figure windows, changing the prefix to correspond to the name of one or the other of the figure windows would direct the axes to be drawn in the appropriate window. Writing ax1 = fig.add_subplot(2, 2, 1) assigns the name ax1 to the axes in the upper left quadrant of the figure window.

The ax1.plot(x, y, 'C0') in line 23 directs Python to plot the previously defined x and y arrays onto the axes named ax1. The statement ax2 = fig.add_subplot(2, 2, 2) draws axes in the second, or upper right, quadrant of the figure window. The statement ax3 = fig.add_subplot(2, 1, 2) divides the figure window into 2 rows (first argument) and 1 column (second argument), creates axes in the second or these two sections, and assigns those axes (*i.e.*, that subplot) the name `ax3`. That is, it divides the figure window into 2 halves, top and bottom, and then draws axes in the half number 2 (the third argument), or lower half of the figure window.

You may have noticed in the above code that some of the function calls are a bit different from those used before, so:

```
xlabel('time (ms)')   → set_xlabel('time (ms)')
title('exponential')  → set_title('exponential')
```

etc.

The call `ax2.set_yscale('log')` sets the y-axes in the second plot to be logarithmic, thus creating a semi-log plot. Alternatively, this can be done with a `ax2.semilogy(x, y, 'C2')` call.

Using the prefixes `ax1`, `ax2`, or `ax3`, directs graphical instructions to their respective subplots. By creating and specifying names for the different figure windows and subplots within them, you access the different plot windows more efficiently. For example, the following code makes four identical subplots in a single figure window using a `for` loop (see §5.2.1).

```
In [1]: fig = figure()

In [2]: ax1 = fig.add_subplot(221)

In [3]: ax2 = fig.add_subplot(222)

In [4]: ax3 = fig.add_subplot(223)

In [5]: ax4 = fig.add_subplot(224)

In [6]: for ax in [ax1, ax2, ax3, ax4]:
   ...:     ax.plot([3,5,8],[6,3,1])

In [7]: fig.show()
```

6.4.1 An alternative syntax for a grid of plots

The syntax introduced above for defining a `Figure` window and opening *a grid* of several subplots can be a bit cumbersome, so an alternative more compact syntax has been developed. We illustrate its use with the program below:

Code: chapter6/programs/multiplePlotsGrid.py

```
1  import numpy as np
2  import matplotlib.pyplot as plt
3
```

Plotting

```python
x = np.linspace(-2*np.pi, 2*np.pi, 200)
sin, cos, tan = np.sin(x), np.cos(x), np.tan(x)
csc, sec, cot = 1.0/sin, 1.0/cos, 1.0/tan

plt.close('all')   # Closes all open figure windows
fig, ax = plt.subplots(2, 3, figsize=(9.5, 6),
                       sharex=True, sharey=True)
ax[0, 0].plot(x, sin, color='red')
ax[0, 1].plot(x, cos, color='orange')
ax[0, 2].plot(x, np.ma.masked_where(np.abs(tan) > 20., tan),
              color='yellow')
ax[1, 0].plot(x, np.ma.masked_where(np.abs(csc) > 20., csc),
              color='green')
ax[1, 1].plot(x, np.ma.masked_where(np.abs(sec) > 20., sec),
              color='blue')
ax[1, 2].plot(x, np.ma.masked_where(np.abs(cot) > 20., cot),
              color='violet')
ax[0, 0].set_xlim(-2*np.pi, 2*np.pi)
ax[0, 0].set_ylim(-5, 5)
ax[0, 0].set_xticks(np.pi*np.array([-2, -1, 0, 1, 2]))
ax[0, 0].set_xticklabels(['-2$\pi$', '-$\pi$', '0',
                          '$\pi$', '2$\pi$'])

ax[0, 2].patch.set_facecolor('lightgray')

ylab = [['sin', 'cos', 'tan'], ['csc', 'sec', 'cot']]
for i in range(2):
    for j in range(3):
        ax[i, j].axhline(color='gray', zorder=-1)
        ax[i, j].set_ylabel(ylab[i][j])

fig.savefig('figures/multiplePlotsGrid.pdf')
fig.show()
fig.canvas.manager.window.raise_()   # fig to front
```

This program generates a 2-row × 3-column grid of plots, as shown in Fig. 6.12, using the function subplots. The first two arguments of subplots specify, respectively, the number of rows and columns in the plot grid. The other arguments are optional; we will return to them after discussing the output of the function subplots.

The output of subplots is a two-element list, which we name fig and ax. The first element fig is the name given to the figure object that contains all of the subplots. The second element ax is the name given to a 2 × 3 list of axes objects, one entry for each subplot. These subplots are indexed as you might expect: ax[0, 0], ax[0, 1], ax[0, 2], ...

Returning to the arguments of subplots, the first keyword argu-

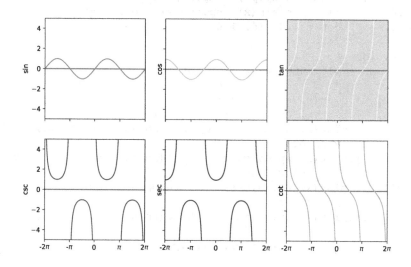

Figure 6.12   Grid of plots.

ment `figsize` sets the overall size of the figure window. The next keyword argument, `sharex=True`, instructs matplotlib to create identical x axes for all six subplots; `sharey=True` does the same for the y axes. Thus, in lines 21 and 22, when the limits are set for the x and y axes only for the first subplot, `[0, 0]`, these instructions are applied to all six subplots because the keyword arguments instruct matplotlib to make all the x axes the same and all the y axes the same. This also applies even to the tick placement and labels, which are set in lines 23–25.

You may have also noticed in lines 24 and 25 that matplotlib can print Greek letters, in this case the letter π. Indeed, matplotlib can output the Greek alphabet as well as virtually any kind of mathematical equations you can imagine using the LaTeX typesetting system. The LaTeX string is enclosed by $ symbols, which are inside of quotes (double or single) because it's a string. The LaTeX capabilities of matplotlib are discussed in §6.6.

By contrast, the subplot background is set to `'lightgray'` only in plot `[0, 2]` in line 27.

The nested `for` loops in lines 30–33 place a gray line at $y = 0$ and labels the y-axis in each subplot.

In line 37 we have added the function call

```
fig.canvas.manager.window.raise_()
```

in order to make the figure display in front of all the other windows on the computer. This is not necessary on all systems; on some the figure window always displays in front. If that's not the case on your installation, you may wish to add this line.

Finally, we note that writing

```
fig, ax = plt.subplots()
```

without any arguments in the `subplots` function opens a figure window with a single subplot. It is equivalent to

```
fig = plt.figure()
ax = fig.add_subplot(111)
```

It's a handy way to save a line of code.

6.5 Plots with multiple axes

Plotting two different quantities that share a common independent variable on the same graph can be a compelling way to compare and visualize data. Figure 6.13 shows an example of such a plot, where the blue curve is linked to the left blue y-axis and the red data points are linked to the right red y-axis. The code below shows how this can be done using matplotlib using the function `twinx()`.

Code: chapter6/programs/twoAxes.py

```python
import numpy as np
import matplotlib.pyplot as plt

fig, ax1 = plt.subplots(figsize=(7.5, 4.5))
xa = np.linspace(0.01, 6.0, 150)
ya = np.sin(np.pi*xa)/xa
ax1.plot(xa, ya, '-C0')
ax1.set_xlabel('x (micrometers)')
# Make y-axis label, ticks and numbers match line color.
ax1.set_ylabel('oscillate', color='C0')
ax1.tick_params('y', colors='C0')

ax2 = ax1.twinx()   # use same x-axis for a 2nd (right) y-axis
xb = np.arange(0.3, 6.0, 0.3)
yb = np.exp(-xb*xb/9.0)
ax2.plot(xb, yb, 'oC3')
ax2.set_ylabel('decay', color='C3')   # axis label
```

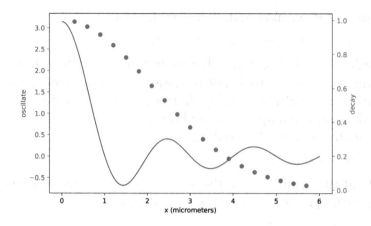

Figure 6.13   Figure with two *y* axes.

```
18  ax2.tick_params('y', colors='C3')      # ticks & numbers
19
20  fig.tight_layout()
21  plt.savefig('figures/twoAxes.pdf')
22  plt.show()
```

After plotting the first set of data using the axes `ax1`, calling `twinx()` instructs matplotlib to use the same *x*-axis for a second *x-y* set of data, which we set up with a new set of axes `ax2`. The `set_ylabel` and `tick_parameters` functions are used to harmonize the colors of the *y*-axes with the different data sets.

There is an equivalent function `twiny()` that allows two sets of data to share a common *y*-axis and then have separate (top and bottom) *x*-axes.

6.6   Mathematics and Greek symbols

matplotlib can display mathematical formulas, Greek letters, and mathematical symbols using a math rendering module known as *mathtext*. The mathtext module parses a subset of Donald Knuth's TeX mathematical typesetting language, and provides basic mathematical typesetting without any software other than matplotlib.

If, in addition, you have TeX (and/or LaTeX) as a separate stand-alone program (such as MacTex [TexShop] or MiKTeX), then you can do even more. In what follows we will assume that you are using the

Plotting

native matplotlib mathtext, but will make a few comments applicable to those who have a separate installation of LaTeX.

matplotlib's mathtext can display Greek letters and mathematical symbols using the syntax of TeX. If you are familiar with TeX or LaTeX, you have hardly anything to learn. Even if you are not familiar with them, the syntax is simple enough that you can employ it in most cases without too much effort.

You designate text as mathtext by placing dollar signs ($) in a text string at the beginning and end of any part of the string that you want to be rendered as math text. You should also use raw strings in most cases, which means you should precede the quotes of a string with the letter r. For example, the following commands produces a plot with the title "$\pi > 3$."

```
In [1]: plot([1, 2, 3, 2, 3, 4, 3, 4, 5])
Out[1]: [<matplotlib.lines.Line2D at 0x11d5c4780>]

In [2]: title(r'$\pi > 3$')
Out[2]: <matplotlib.text.Text at 0x11d59f390>
```

Where the matplotlib function normally takes a string as input, you simply input the mathtext string. Note the r before the string in the title argument and the dollar signs ($) inside the quotes at the beginning and end of the text you wish to render as math text.

Subscripts and superscripts are designated using the underline "_" and caret "^" characters, respectively. Multiple characters to be included together in a subscript or superscript should be enclosed in a pair of curly braces {...}. All of this and more is illustrated in the plot shown in Fig. 6.14, which is produced by the Python code below.

Code: chapter6/programs/mplLatexDemo.py

```
1  import numpy as np
2  import matplotlib.pyplot as plt
3
4
5  def f0(t, omega, gamma, tau):
6      wt = omega*t
7      f1 = np.sin(wt) + (np.cos(wt)-1.0)/wt
8      f2 = 1.0+(gamma/omega)*f1
9      return np.exp(-t*f2/tau)
10
11
12 omega = 12.0
13 gamma = 8.0
14 tau = 1.0
```

```
15  t = np.linspace(0.01, 10.0, 500)
16  f = f0(t, omega, gamma, tau)
17
18  plt.rc('mathtext', fontset='stix')      # Use with mathtext
19  # plt.rc('text', usetex=True)            # Use with Latex
20  # plt.rc('font', family='serif')         # Use with Latex
21
22  fig, ax = plt.subplots(figsize=(7.5, 4.5))
23  ax.plot(t, f, color='C0')
24  ax.set_ylabel(r'$f_0(t)$', fontsize=14)
25  ax.set_xlabel(r'$t/\tau\quad\rm(ms)}$', fontsize=14)
26  ax.text(0.45, 0.95,
27          r'$\Gamma(z)=\int_0^\infty x^{z-1}e^{-x}dx$',
28          fontsize=16, ha='right', va='top',
29          transform=ax.transAxes)
30  ax.text(0.45, 0.75,
31          r'$e^x=\sum_{n=0}^\infty\frac{x^n}{n!}$',
32          fontsize=16, ha='right', va='top',
33          transform=ax.transAxes)
34  ax.text(0.45, 0.55,
35          r'$\zeta(z)=\prod_{k=0}^\infty \frac{1}{1-p_k^{-z}}$',
36          fontsize=16, ha='left', va='top',
37          transform=ax.transAxes)
38  ax.text(0.95, 0.80,
39          r'$\omega={0:0.1f},\;\gamma={1:0.1f},\;\tau={2:0.1f}$'
40          .format(omega, gamma, tau),
41          fontsize=14, ha='right', va='top',
42          transform=ax.transAxes)
43  ax.text(0.85, 0.35,
44          r'$e=\lim_{n\to\infty}\left(1+\frac{1}{n}\right)^n$',
45          fontsize=14, ha='right', va='top',
46          transform=ax.transAxes)
47
48  fig.tight_layout()
49  fig.savefig('./figures/mplLatexDemo.pdf')
50  fig.show()
```

Line 18 sets the font to be used by mathtext, in this case `stix`. If you omit such a statement, mathtext uses its default font `dejavusans`. Other options include `dejavuserif`, `cm` (Computer Modern), `stix`, and `stixsans`. Try them out!

Let's see what LaTeX can do. First, we see that the *y*-axis label in Fig. 6.14 is $f_0(t)$, which has a subscript that was formatted using mathtext in line 24. The *x*-axis label is also typeset using mathtext, but the effects are more subtle in this case. The variable *t* is italicized, as is proper for a mathematical variable, but the units, (ms), are not, which is also standard practice. The math mode italics are turned off with

Plotting

the \rm (Roman) switch, which acts on text until the next closing curly brace (}).

Lines 26–29 provide the code to produce the expression for the Gamma function $\Gamma(z)$, lines 30–33 produce the expression for the Taylor series for the exponential function, and lines 34–37 produce the product that gives the zeta function. Lines 38–42 provide the code that produces the expressions for ω, γ, and τ. Lines 43–46 provide the code that produces the limit expression for the natural logarithm base e. The mathtext (TEX) code that produces the four typeset equations is contained in strings in lines 27, 31, 35, 39, and 44. The strings begin and end with $ symbols, which activates and deactivates mathtext's math mode.

Special commands and symbols begin with a backslash in mathtext, which follows the convention of TEX. However, Python strings also use backslashes for certain formatting commands, such as \t for TAB or \n for a new line. The r preceding the strings in lines 27, 31, 35, 39, and 44 makes those strings *raw* strings, which turns off Python's special backslash formatting commands so that backslash commands are interpreted as mathtext.

Table 6.2 provides the LATEX (mathtext) codes for Greek letters; Table 6.3 gives the code for miscellaneous mathematical expressions.

The mathtext codes are the same as LATEX codes but are sometimes rendered slightly differently from what you might be used to if you

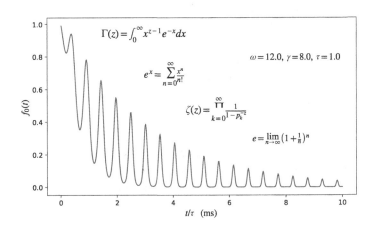

Figure 6.14 Plot using matplotlib's mathtext for Greek letters and mathematical symbols.

α	\alpha	β	\beta	γ	\gamma	δ	\delta
ϵ	\epsilon	ε	\varepsilon	ζ	\zeta	η	\eta
θ	\theta	ι	\iota	κ	\kappa	λ	\lambda
μ	\mu	ν	\nu	ξ	\xi	π	\pi
ρ	\rho	ϱ	\varrho	σ	\sigma	τ	\tau
υ	\upsilon	ϕ	\phi	φ	\varphi	χ	\chi
ψ	\psi	ω	\omega	Γ	\Gamma	Δ	\Delta
Θ	\Theta	Λ	\Lambda	Ξ	\Xi	Π	\Pi
Σ	\Sigma	Υ	\Upsilon	Φ	\Phi	Ψ	\Psi
Ω	\Omega						

Table 6.2 LaTeX (mathtext) codes for Greek letters.

use LaTeX. For example, the code `$\cos\theta$` produces $\cos\theta$ in LaTeX, but produces $\cos\theta$ using matplotlib's mathtext: there is too little space between cos and θ in the mathtext expression. For this reason, you may want to insert an extra bit of space in your mathtext code where you wouldn't normally need to do so using LaTeX. Spaces of varying length can be inserted using `\,`, `\:` `\;` and `\ ` for shorter spaces (of increasing length); `\quad` and `\qquad` provide longer spaces equal to one and two character widths, respectively.

While extra spacing in rendered mathtext equations matters, extra spaces in mathtext *code* makes no difference in the output. In fact, spaces in the code are not needed if the meaning of the code is unambiguous. Thus, `$\cos \, \theta$` and `$\cos\,\theta$` produce exactly the same output, namely $\cos\theta$.

matplotlib can produce even more beautiful mathematical expressions if you have a separate stand-alone version of TeX (and/or LaTeX) on your computer. In this case, you have access to a broader selection of fonts—all the fonts you have installed with your TeX installation. You also can write LaTeX code using the `\displaystyle` switch, which produces more nicely proportioned expressions. To do this, each expression in lines 27, 31, 35, and 44 should have `\displaystyle` prepended to each mathtext string. You would also uncomment lines 19–20 and comment out line 18. The result is shown in Fig. 6.15.[3]

---

[3] For LaTeX experts, matplotlib's mathtext does not recognize TeX's `$$...$$` syntax for displayed equations, but you can get the same result with the `\displaystyle` switch.

Plotting

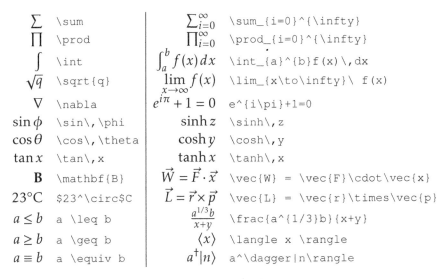

Table 6.3 Mathtext (LaTeX) codes for miscellaneous mathematical expressions. Search "latex math symbols" on the internet for more extensive lists.

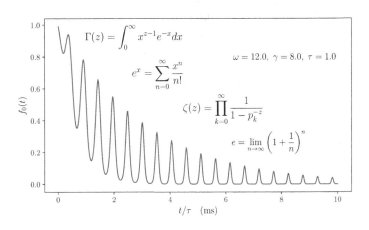

Figure 6.15 Plot using LaTeX for Greek letters and mathematical symbols.

6.7 The Structure of matplotlib: OOP and All That

In this section, we provide an overview of the logical structure of matplotlib. On a first pass, you can certainly skip this section, but you may find it useful for a variety of reasons. First, it will help you better understand the matplotlib syntax we introduced in §6.4. Second,

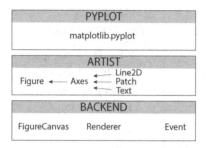

Figure 6.16 matplotlib software layers, from the low-level backend to the high level PyPlot.

it should improve your ability to read and understand the official online documentation as well as other online resources such as our favorite, stack**overflow**.[4] The writing in these and other web resources is replete with jargon and ideas than can be frustratingly obscure to a novice. Much of it is the jargon of object-oriented programming. Other parts pertain to the jargon of graphical user interfaces, or GUIs. In this section, we introduce the basic structure of matplotlib and explain its lexicon.

matplotlib is a Python module for generating graphical output to your computer screen or to a computer file. Fundamentally, its job is to translate Python scripts into graphical instructions that your computer can understand. It does this using two different layers of software, the *backend* layer and the *artist* layer. To these two layers it adds a scripting layer, PyPlot, which we have met already (`import matplotlib.pyplot as plt`). PyPlot is a convenience layer, and not really necessary, but it facilitates rapid scripting, and aids with portability. As you shall see, for most programming we advocate using a hybrid of the scripting and artist layers. Figure 6.16 portrays the matplotlib software hierarchy.

6.7.1 The backend layer

For the sake of being concrete, we'll start by considering the task of creating a figure on your computer screen. matplotlib must be able

---

[4]https://stackoverflow.com/tags/matplotlib/info

Plotting

to generate graphics using different computer platforms: Linux, Microsoft Windows, and macOS. Of course, we want matplotlib to do this in a way that is transportable from one platform to another and transparent to you the user. Ideally, the Python matplotlib code you write on your PC should work without modification on your friend's Mac or Linux computer.

To accomplish this, matplotlib uses open source cross-platform software toolkits written by third parties. There are a number of different toolkits available. Most of these toolkits can write graphics to computer screens on different platforms, including Windows, macOS, and Linux. They are written mostly in C++, and are very efficient and powerful. To harness their power and versatility, matplotlib provides a number of Python "wrappers"—Python functions—that call the C++ functions of these toolkits to send graphics instructions to your computer.[5] matplotlib calls these wrappers *backends*.

Several backends have been written for matplotlib. They fall into two categories: those written for output to files—hardcopy backends—and those for output to a computer screen—user interface backends, also known as interactive backends.

For output to a computer screen, the qt5Agg backend is among the most versatile and widely used, so we will use it as an example of what an interactive backend does.[6] The qt5Agg backend is made from two C++ toolkits: Qt[7] and Agg.[8] Qt can define windows on your computer screen, create buttons, scrollbars, and other *widgets*, that is to say, elements of a GUI. It can also process *events*, actions like clicking a mouse, pressing a keyboard key, moving a scroll bar, or pressing a button. This includes processing events generated by the computer, rather than the user, like an alarm clock going off or a process finishing. The Agg toolkit is a rendering program that produces pixel images in memory from vectorial data. For example, you provide Agg with the equation for a circle and it determines which pixels to

---

[5] A Python "wrapper" provides a Python interface to software written in a different computer language, such as C++ or Fortran. You call a Python function and it calls a C++ or Fortran program.

[6] Your installation may use the qt4Agg backend instead of the qt5Agg backend. In that case, just substitute qt4Agg for qt5Agg in what follows.

[7] https://en.wikipedia.org/wiki/Qt_(software)

[8] https://en.wikipedia.org/wiki/Anti-Grain_Geometry

activate on the computer screen. It employs advanced rendering techniques like *anti-aliasing*[9] to produce faithful high-resolution graphics.

The job of a matplotlib backend is to provide a Python interface to the functionality of the underlying C++ (or other language) toolkits, which for qt5Agg are qt5 and Agg. The Python-facing parts of all matplotlib backends have three basic classes:

FigureCanvas defines the canvas—a figure window or a graphics file—and transfers the output from the Renderer onto this canvas. It also translates Qt events into the matplotlib Event framework (see below).

Renderer does the drawing. Basically, it connects matplotlib to the Agg library described above.

Event handles user inputs such as keyboard and mouse events for matplotlib.

Besides the qt5Agg backend, there are several other commonly used interactive backends: TkAgg, GTK3Agg, GTK3Cairo, WXAgg, and macOSX to name a few. Why all the different backends? Part of the reason is historical. Early on in the development of matplotlib, many toolkits worked on only one platform, meaning that a separate backend had to be developed for each one. As time has passed, most toolkits became cross-platform. As better cross-platform graphical tools, generally written in C++, were developed, programmers in the Python world wanted access to their functionality. Hence the different backends.

As mentioned earlier, there are also hardcopy backends that only produce graphical output to files. These include Agg (which we already met as part of qt5Agg), PDF (to produce the Adobe portable document format), SVG (scalable vector graphics), and Cairo (png, ps, pdf, and svg).

In the end, the idea of a matplotlib backend is to provide the software machinery for setting up a canvas to draw on and the low-level tools for creating graphical output: plots and images. The drawing tools of the backend layer, while sufficient for producing any output you might want, work at too low of a level to be useful for everyday programming. For this, matplotlib provides another layer, the artist

---

[9]See https://en.wikipedia.org/wiki/Spatial_anti-aliasing.

Plotting

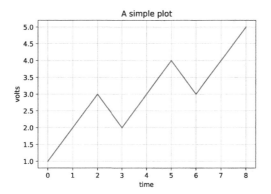

Figure 6.17 Plotting without PyPlot using the pure OO-interface.

layer, which provides the software tools you will use for creating and managing graphical output.

6.7.2 The artist layer

The artist layer consists of a hierarchy of Python classes that facilitate creating a figure and embellishing it with any and all of the features we might desire: axes, data points, curves, axes labels, legends, titles, annotations, and everything else. The first step is to create a figure and place it on a canvas (figure window, whether on a computer screen or a computer file). To the figure, we add axes, data, labels, *etc.* When we have everything the way we want it, we send it to the screen to display and/or save it to a file.

To make this concrete, consider the program below that creates the plot shown in Fig. 6.17.

Code: chapter6/programs/oopTest.py

```
from matplotlib.backends.backend_qt5agg \
    import FigureCanvasQTAgg as FigureCanvas
from matplotlib.figure import Figure

fig = Figure(figsize=(6, 4))
canvas = FigureCanvas(fig)
ax = fig.add_subplot(111)
ax.plot([1, 2, 3, 2, 3, 4, 3, 4, 5])
ax.set_title('A simple plot')
ax.grid(True)
ax.set_xlabel('time')
```

```
12  ax.set_ylabel('volts')
13  canvas.print_figure('figures/oopTest.pdf')
14  canvas.show()
```

First, we import the `FigureCanvas` from the qt5Agg backend.[10] Then we import `Figure` from `matplotlib.figure`. After finishing our imports, the first step is to define and name (`fig`) the `Figure` object that will serve as a container for our plot. The next step is to attach the figure to an instance of `FigureCanvas`, which we name `canvas`, from the `qt5Agg` backend. This places the canvas on the screen and connects all the matplotlib routines to the Qt and Agg C++ routines that write to the screen. Next, in line 7, we create a set of axes on our figure, making a single subplot that takes up the entire frame of the figure. In fact, lines 7–12 should be completely familiar to you, as they use the syntax introduced in §6.4. Finally, we write the plot to a file (line 13) and to the screen (line 14) from the canvas, which makes sense, because that's what connects the matplotlib routines to the hardware.

The code in this example is native matplotlib object-oriented (OO) code. In its purest form, it's the way matplotlib code is supposed to be written. The code is entirely transportable from one computer to another irrespective of the operating system, so long as the qt5Agg backend has been included in the local machine's Python installation. In this case, the output should look the same, whether it's on the screen or in a file, on Microsoft Windows, macOS, or any of the different flavors of Linux. By the way, if another machine does not have the qt5Agg backend installed, you can simply change lines 1 and 2 to a different backend that is installed, and the program should work as expected, with no discernible differences.

Before moving on to the next layer of matplotlib, it's useful to introduce some matplotlib terminology—jargon—for describing the artist layer. All of the routines called in lines 5 and 7–12 are part of the Artist module and the eponymous Artist class of matplotlib. Collectively and individually, all the routines that get attached to `fig` and its descendant `ax` are known as *Artists*: `add_subplot`, `plot`, `title`, `grid`, `xlabel`, *etc*. Artists are those matplotlib objects that draw on the canvas, including `figure`. You will see the term Artist employed liberally in online documentation and commentary on matplotlib. It's simply

---

[10]Lines 1–2 of can be shortened to read:
`from matplotlib.backends.backend_qt5Agg import FigureCanvas`

Plotting

a part of the matplotlib lexicon, along with backend, PyPlot, and the yet to be mentioned, PyLab.

6.7.3 The PyPlot (scripting) layer

For newcomers to Python, learning about backends and the OOP syntax can create a barrier to code development, particularly engineers and scientists who come to Python after having first been exposed to MATLAB®. This is why the PyPlot module was developed, to provide a simpler and more familiar interface to those coming from software packages like MATLAB®, or simply for those who are new to Python programming.

Consider, for example, our first plotting script of a sine function, which we reproduce here from page 102:

Code: chapter6/programs/sineFunctionPlot33.py

```
import numpy as np
import matplotlib.pyplot as plt
x = np.linspace(0, 4.*np.pi, 33)
y = np.sin(x)
plt.plot(x, y)
plt.show()
```

After importing NumPy and PyPlot, we simply define the x-y arrays in two lines, make a plot in the next line, and then display it in the last line: the epitome of simplicity.

PyPlot and backends

There is no mention of a backend in this syntax. Nevertheless, one must be used so that matplotlib can communicate with the computer. So how does a backend get loaded using PyPlot? And which backend? First, a backend is loaded when PyPlot is imported. Second, which backend is loaded is determined by how you launch IPython. If you launch it from a terminal or from a Jupyter notebook, the default backend is set by a matplotlib configuration file named `matplotlibrc` on your computer. The file was put on your computer, perhaps without your knowledge, when matplotlib was installed, for example using the Anaconda or Enthought installations of Python. You can find out where this file is from an IPython prompt by typing

```
In [1]: import matplotlib

In [2]: matplotlib.matplotlib_fname()
```

```
Out[2]: '/Users/pine/.matplotlib/matplotlibrc'
```

Alternatively, if you are using a Python IDE like Spyder, the backend that is used is set in a Preferences menu and loaded when the IDE is launched. In any case, you can find out which backend was installed in your IPython shell using a IPython magic command

```
In [3]: %matplotlib
Out[3]: Using matplotlib backend: Qt5Agg
```

Thus, using PyPlot frees you up from thinking about the backend—how matplotlib is connected to your computer's hardware—when writing matplotlib routines. That's a good thing.

PyPlot's "state-machine" environment

For very *simple plots*, you can proceed using the syntax introduced in §6.2 and §6.3. In this syntax, the only matplotlib prefix used is `plt`. You can do most of the things you want to do with matplotlib working in this way. You can make single plots or plots with multiple subplots, as demonstrated in §6.4.1.

When working in this mode with PyPlot, we are working in a "state-machine" environment, meaning that where on a canvas PyPlot adds features (*i.e.*, Artists) depends on the state of the program. For example, in the program starting on page 114 that produces the plot in Fig. 6.9, you make first subplot by including `plt.subplot(2, 1, 1)` in the program. Everything that comes after that statement affects that subplot, until the program comes to the line `plt.subplot(2, 1, 2)`, which opens up a second subplot. After that statement, everything affects the second subplot. You've changed the "state" of the "machine" (*i.e.*, the FigureCanvas object). Which subplot is affected by commands in the program depends on which state the machine is in.

PyPlot's hybrid OOP environment

Contrast operating in this state-machine mode with the syntax introduced in §6.4. In those programs, we load PyPlot, but then employ the OOP syntax for creating figures and making subplots. Each subplot object is referenced by a different object name, *e.g.*, `ax1`, `ax2`, or `ax[i, j]`, where `i` and `j` cycle through different values. These object names identify the target of a function, not the state of the machine. It's a more powerful way of programming and provides more versatility in

Plotting

writing code. For example, in making the plots shown in Fig. 6.12, we could conveniently combine the labeling of the y-axes by cycling through object names of each subplot towards the end of the program (see page 123). This would be hard to do with the state-machine approach.

In general, most programming with matplotlib should be done using this hybrid (but basically OOP) approach introduced in §6.4. The state-machine approach we employed in earlier sections should be reserved for short snippets of code.

6.8 Contour and Vector Field Plots

matplotlib has extensive tools for creating and annotating two-dimensional contour plots and vector field plots. A *contour plot* is used to visualize two-dimensional scalar functions, such as the electric potential $V(x,y)$ or elevations $h(x,y)$ over some physical terrain. Vector field plots come in different varieties. There are field line plots, which in some contexts are called *streamline plots*, that show the direction of a vector field over some 2D (x,y) range. There are also *quiver plots*, which consist essentially of a 2D grid of arrows, that give the direction and magnitude of a vector field over some 2D (x,y) range.

6.8.1 Making a 2D grid of points

When plotting a function $f(x)$ of a single variable, the first step is usually to create a one-dimensional x array of points, and then to evaluate and plot the function $f(x)$ at those points, often drawing lines between the points to create a continuous curve. Similarly, when making a two-dimensional plot, we usually need to make a two-dimensional x-y array of points, and then to evaluate and plot the function $f(x,y)$, be it a scaler or vector function, at those points, perhaps with continuous curves to indicate the value of the function over the 2D surface.

Thus, instead of having a line of evenly spaced x points, we need a *grid* of evenly spaced x-y points. Fortunately, NumPy has a function `np.meshgrid` for doing just that. The procedure is first to make an x-array at even intervals over the range of x to be covered, and then to do the same for y. These two one-dimensional arrays are used as input to the `np.meshgrid` function, which makes a two-dimensional mesh. Here is how it works:

```
In [1]: x = linspace(-1, 1, 5)

In [2]: x
Out[2]: array([-1. , -0.5,  0. ,  0.5,  1. ])

In [3]: y = linspace(2, 6, 5)

In [4]: y
Out[4]: array([ 2.,  3.,  4.,  5.,  6.])

In [5]: X, Y = np.meshgrid(x, y)

In [6]: X
Out[6]:
array([[-1. , -0.5,  0. ,  0.5,  1. ],
       [-1. , -0.5,  0. ,  0.5,  1. ],
       [-1. , -0.5,  0. ,  0.5,  1. ],
       [-1. , -0.5,  0. ,  0.5,  1. ],
       [-1. , -0.5,  0. ,  0.5,  1. ]])

In [7]: Y
Out[7]:
array([[ 2.,  2.,  2.,  2.,  2.],
       [ 3.,  3.,  3.,  3.,  3.],
       [ 4.,  4.,  4.,  4.,  4.],
       [ 5.,  5.,  5.,  5.,  5.],
       [ 6.,  6.,  6.,  6.,  6.]])

In [7]: plot(X, Y, 'o')
```

The output of plot(X, Y, 'o') is a 2D grid of points, as shown in Fig. 6.18. matplotlib's functions for making contour plots and vector field plots generally use the output of gridmesh as the 2D input for the functions to be plotted.

6.8.2 Contour plots

The principal matplotlib routines for creating contour plots are contour and contourf. Sometimes you would like to make a contour plot of a function of two variables; other times you may wish to make a contour plot of some data you have. Of the two, making a contour plot of a function is simpler, which is all we cover here.

Contour plots of functions

Figure 6.19 shows four different contour plots. All were produced using contour except the upper left plot which was produced using

Plotting

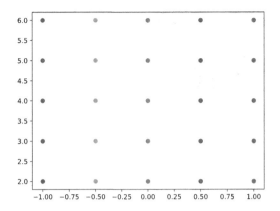

Figure 6.18   Point pattern produced by np.gridmesh(X, Y).

contourf. All plot the same function, which is the sum of a pair of Gaussians, one positive and the other negative:

$$f(x,y) = 2e^{-\frac{1}{2}[(x-2)^2+(y-1)^2]} - 3e^{-2[(x-1)^2+(y-2)^2]} \qquad (6.1)$$

The code that produces Fig. 6.19 is given below.

Code: chapter6/programs/contour4.py

```
import numpy as np
import matplotlib.pyplot as plt
import matplotlib.cm as cm       # color maps
import matplotlib


def pmgauss(x, y):
    r1 = (x-1)**2 + (y-2)**2
    r2 = (x-3)**2 + (y-1)**2
    return 2*np.exp(-0.5*r1) - 3*np.exp(-2*r2)


a, b = 4, 3

x = np.linspace(0, a, 60)
y = np.linspace(0, b, 45)

X, Y = np.meshgrid(x, y)
Z = pmgauss(X, Y)

```

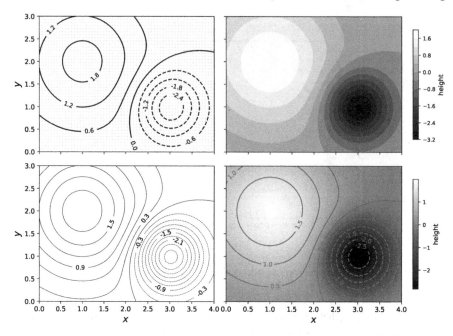

Figure 6.19   Contour plots.

```
21  fig, ax = plt.subplots(2, 2, figsize=(9.4, 6.5),
22                         sharex=True, sharey=True,
23                         gridspec_kw={'width_ratios': [4, 5]})
24
25  CS0 = ax[0, 0].contour(X, Y, Z, 8, colors='k')
26  ax[0, 0].clabel(CS0, fontsize=9, fmt='%0.1f')
27  matplotlib.rcParams['contour.negative_linestyle'] = 'dashed'
28  ax[0, 0].plot(X, Y, 'o', ms=1, color='lightgray', zorder=-1)
29
30  CS1 = ax[0, 1].contourf(X, Y, Z, 12, cmap=cm.gray, zorder=0)
31  cbar1 = fig.colorbar(CS1, shrink=0.8, ax=ax[0, 1])
32  cbar1.set_label(label='height', fontsize=10)
33  plt.setp(cbar1.ax.yaxis.get_ticklabels(), fontsize=8)
34
35  lev2 = np.arange(-3, 2, 0.3)
36  CS2 = ax[1, 0].contour(X, Y, Z, levels=lev2, colors='k',
37                         linewidths=0.5)
38  ax[1, 0].clabel(CS2, lev2[1::2], fontsize=9, fmt='%0.1f')
39
40  CS3 = ax[1, 1].contour(X, Y, Z, 10, colors='gray')
41  ax[1, 1].clabel(CS3, fontsize=9, fmt='%0.1f')
42  im = ax[1, 1].imshow(Z, interpolation='bilinear',
43                       origin='lower', cmap=cm.gray,
```

Plotting

```
44                      extent=(0, a, 0, b))
45      cbar2 = fig.colorbar(im, shrink=0.8, ax=ax[1, 1])
46      cbar2.set_label(label='height', fontsize=10)
47      plt.setp(cbar2.ax.yaxis.get_ticklabels(), fontsize=8)
48
49      for i in range(2):
50          ax[1, i].set_xlabel(r'$x$', fontsize=14)
51          ax[i, 0].set_ylabel(r'$y$', fontsize=14)
52          for j in range(2):
53              ax[i, j].set_aspect('equal')
54              ax[i, j].set_xlim(0, a)
55              ax[i, j].set_ylim(0, b)
56      plt.subplots_adjust(left=0.06, bottom=0.07, right=0.99,
57                          top=0.99, wspace=0.06, hspace=0.09)
```

After defining the function to be plotted in lines 7–10, the next step is to create the *x-y* array of points at which the function will be evaluated using `np.meshgrid`. We use `np.linspace` rather than `np.arange` to define the extent of the *x-y* mesh because we want the *x* range to go precisely from 0 to `a=4` and the *y* range to go precisely from 0 to `b=3`. We use `np.linspace` for two reasons. First, if we use `np.arange`, the array of data points does not include the upper bound, while `np.linspace` does. This is important for producing the grayscale (or color) background that extends all the way to the upper limits of the *x-y* ranges in the upper-right plot, produced by `contourf`, of Fig. 6.19. Second, to produce smooth-looking contours, one generally needs about 40–200 points in each direction across the plot, irrespective of the absolute magnitude of the numbers being plotted. The number of points is directly specified by `np.linspace` but must be calculated for `np.arange`. We follow the convention that the `meshgrid` variables are capitalized, which seems to be a standard followed by many programmers. It's certainly not necessary.

The upper-left contour plot takes the X-Y 2D arrays made using `gridspec` as its first two arguments and Z as its third argument. The third argument tells `contour` to make approximately 5 different levels in Z. We give the `contour` object a name, as it is needed by the `clabel` call in the next line, which sets the font size and the format of the numbers that label the contours. The line style of the negative contours is set globally to be "dashed" by a call to matplotlib's `rcparams`. We also plot the location of the X-Y grid created by `gridspec` just for the sake of illustrating its function; normally these would not be plotted.

The upper-right contour plot is made using `contourf` with 12 different z layers indicated by the different gray levels. The gray color scheme is set by the keyword argument `cmap`, which here is set to the `matplotlib.cm` color scheme `cm.gray`. Other color schemes can be found in the matplotlib documentation by an internet search on "matplotlib choosing colormaps." The color bar legend on the right is created by the `colorbar` method, which is attached to `fig`. It is associated with the upper right plot by the name `CS1` of the `contourf` method *and* by the keyword argument `ax=ax[0, 1]`. Its size relative to the plot is determined by the `shrink` keyword. The font size of the color bar label is set using the generic set property method `setp` using a somewhat arcane but compact syntax.

For the lower-left contour plot `CS2`, we manually specify the levels of the contours with the keyword argument `levels=lev2`. We specify that only every other contour will be labeled numerically with `lev2[1::2]` as the second argument of the `clabel` call in line 38; `lev2[0::2]` would also label every other contour, but the even ones instead of the odd ones.

The lower-right contour plot `CS3` has 10 contour levels and a continuously varying grayscale background created using `imshow`. The `imshow` method uses only the z array to determine the gray levels. The *x-y* extent of the grayscale background is determined by the keyword argment `extent`. By default, `imshow` uses the upper-left corner as its origin. We override the default using the `imshow` keyword argument `origin='lower'` so that the grayscale is consistent with the data. The keyword argument `iterpolation` tells `imshow` how to interpolate the grayscale between different z levels.

6.8.3 Streamline plots

matplotlib can also make streamline plots, which are sometimes called field line plots. The matplotlib function call to make such plots is `streamplot`, and its use is illustrated in Fig. 6.20 to plot the streamlines of the velocity field of a viscous liquid around a sphere falling through it at constant velocity u. The left plot is in the reference frame of the falling sphere and the right plot is in the laboratory frame where the liquid very far from the sphere is at rest. The program that produces Fig. 6.20 is given below.

Code: chapter6/programs/stokesFlowStream.py

```
1  import numpy as np
```

Plotting

```python
import matplotlib.pyplot as plt
from matplotlib.patches import Circle


def v(u, a, x, z):
    """Return the velocity vector field v = (vx, vy)
    around sphere at r=0."""
    r = np.sqrt(x*x+z*z)
    R = a/r
    RR = R*R
    cs, sn = z/r, x/r
    vr = u * cs * (1.0 - 0.5 * R * (3.0 - RR))
    vtheta = -u * sn * (1.0 - 0.25 * R * (3.0 + RR))
    vx = vr * sn + vtheta * cs
    vz = vr * cs - vtheta * sn
    return vx, vz


# Grid of x, y points
xlim, zlim = 12, 12
nx, nz = 100, 100
x = np.linspace(-xlim, xlim, nx)
z = np.linspace(-zlim, zlim, nz)
X, Z = np.meshgrid(x, z)

# Set particle radius and velocity
a, u = 1.0, 1.0

# Velocity field vector, V=(Vx, Vz) as separate components
Vx, Vz = v(u, a, X, Z)

fig, (ax1, ax2) = plt.subplots(1, 2, figsize=(9, 4.5))

# Plot the streamlines using colormap and arrow style
color = np.log(np.sqrt(Vx*Vx + Vz*Vz))
seedx = np.linspace(-xlim, xlim, 18)   # Seed streamlines
seedz = -zlim * np.ones(len(seedx))    # evenly in far field
seed = np.array([seedx, seedz])
ax1.streamplot(x, z, Vx, Vz, color=color, linewidth=1,
               cmap='afmhot', density=5, arrowstyle='-|>',
               arrowsize=1.0, minlength=0.4,
               start_points=seed.T)
ax2.streamplot(x, z, Vx, Vz-u, color=color, linewidth=1,
               cmap='afmhot', density=5, arrowstyle='-|>',
               arrowsize=1.0, minlength=0.4,
               start_points=seed.T)
for ax in (ax1, ax2):
    # Add filled circle for sphere
    ax.add_patch(Circle((0, 0), a, color='C0', zorder=2))
```

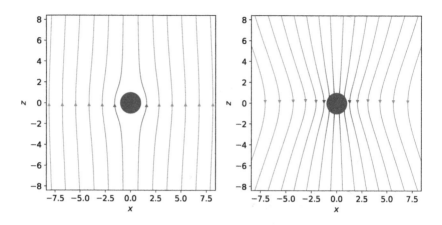

Figure 6.20 Streamlines of flow around a sphere falling in a viscous fluid.

```
51      ax.set_xlabel('$x$')
52      ax.set_ylabel('$z$')
53      ax.set_aspect('equal')
54      ax.set_xlim(-0.7*xlim, 0.7*xlim)
55      ax.set_ylim(-0.7*zlim, 0.7*zlim)
56  fig.tight_layout()
57  fig.savefig('./figures/stokesFlowStream.pdf')
58  fig.show()
```

The program starts by defining a function that calculates the velocity field as a function of the lateral distance x and the vertical distance z. The function is a solution to the Stokes equation, which describes flow in viscous liquids at very low (zero) Reynolds number. The velocity field serves as the primary input into the matplotlib streamplot function.

The next step is to use NumPy's meshgrid program to define the 2D grid of points at which the velocity field will be calculated, just as we did for the contour plots. After setting up the meshgrid arrays X and Z, we call the function we defined v(u, a, X, Z) to calculate the velocity field (line 31).

The streamplot functions are set up in lines 36–39 and called in lines 40–47. Note that for the streamplot function the input x-z coordinate arrays are 1D arrays but the velocity arrays vx-vz are 2D arrays. The arrays seedx and seedx set up the starting points (seeds) for the streamlines. You can leave them out and streamplot will make its own choices based on the values you set for the density and minlength

Plotting

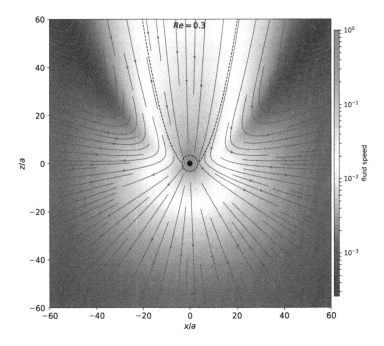

Figure 6.21   Streamlines of flow around a sphere falling in a fluid.

keywords. Here we have chosen them, along with the seed settings, so that all the streamlines are continuous across the plot. The other keywords set the properties for the arrow size and style, the width of the streamlines, and the coloring of the streamlines, in this case according to the speed at a given point.

Let's look at another streamline plot, which illustrates some other possibilities for customizing streamline plots. The plot in Fig. 6.21 shows the streamlines for a faster moving sphere, and makes different choices than the plot above. The code to make this plot, stokesOseenFlow.py, is provided on the next page. The most noticeable difference is the use of the matplotlib function pcolor in lines 58–60 that adds background coloring to the plot keyed to the local speed of the liquid. A logarithmic color scale is used with a logarithmic color bar, which is set up by setting the pcolor keyword norm=LogNorm(vmin=speed.min(), vmax=1) in line 59.

Code: chapter6/programs/stokesOseenFlow.py

```python
import numpy as np
import matplotlib.pyplot as plt
import matplotlib
from matplotlib.patches import Circle
from matplotlib.colors import LogNorm

matplotlib.rcParams.update({'font.size': 12})


def v(u, a, x, z, Re):
    """Return the velocity vector field v = (vx, vy)
    around sphere at r=0."""
    theta = np.arctan2(-x, -z)
    cs, sn = np.cos(theta), np.sin(theta)
    R = a/np.sqrt(x*x+z*z)
    if Re > 0:       # Oseen solution
        ex = np.exp(-0.5*Re*(1.0+cs)/R)
        vr = 0.5 * u * R * \
            (1.5*(1.0-cs)*ex - R*(3.0*(1-ex)/Re - R*cs))
        vtheta = 0.25 * u * R * sn * (3.0*ex - R*R)
    else:            # Stokes solution
        RR = R*R
        vr = 0.5 * u * cs * R * (RR - 3.0)
        vtheta = 0.25 * u * sn * R * (RR + 3.0)
    vx = vr * sn + vtheta * cs
    vz = vr * cs - vtheta * sn
    return vx, vz


def stokesWake(x, Re):
    """Return parabola r[1+cos(theta)]=xi of Stokes wake"""
    z = -0.5 * (1.0/Re - x*x*Re)
    return np.ma.masked_where(x*x+z*z < 1.0/Re**2, z)


# Set particle radius and velocity
a, u = 1.0, 1.0      # normalizes radius & velocity
Re = 0.3             # Reynolds number (depends on viscosity)

# Grid of x, z points
xlim, zlim = 60, 60
nx, nz = 200, 200
x = np.linspace(-xlim, xlim, nx)
z = np.linspace(-zlim, zlim, nz)
X, Z = np.meshgrid(x, z)

# Velocity field vector, v=(Vx, Vz) as separate components
```

Plotting 149

```
48  Vx, Vz = v(u, a, X, Z, Re)
49  R = np.sqrt(X*X+Z*Z)
50  speed = np.sqrt(Vx*Vx+Vz*Vz)
51  speed[R < a] = u   # set particle speed to u
52
53  fig, ax = plt.subplots(figsize=(8, 8))
54
55  # Plot the streamlines with an bwr colormap and arrow style
56  ax.streamplot(x, z, Vx, Vz, linewidth=1, density=[1, 2],
57                arrowstyle='-|>', arrowsize=0.7, color='C0')
58  cntr = ax.pcolor(X, Z, speed,
59                   norm=LogNorm(vmin=speed.min(), vmax=1),
60                   cmap=plt.cm.bwr)
61  if Re > 0:
62      ax.add_patch(Circle((0, 0), 1/Re, color='black',
63                          fill=False, ls='dashed', zorder=2))
64      ax.plot(x, stokesWake(x, Re), color='black', lw=1,
65             ls='dashed', zorder=2)
66  cbar = fig.colorbar(cntr, ax=ax, aspect=50, fraction=0.02,
67                      shrink=0.9, pad=0.01)
68  cbar.set_label(label='fluid speed', fontsize=10)
69  plt.setp(cbar.ax.yaxis.get_ticklabels(), fontsize=10)
70  cbar.set_clim(speed.min(), 1)
71  cbar.draw_all()
72
73  # Add filled circle for sphere
74  ax.add_patch(Circle((0, 0), a, color='black', zorder=2))
75  ax.set_xlabel('$x/a$')
76  ax.set_ylabel('$z/a$')
77  ax.set_aspect(1)
78  ax.set_xlim(-xlim, xlim)
79  ax.set_ylim(-zlim, zlim)
80  ax.text(0.5, 0.99, r"$Re = {0:g}$".format(Re), ha='center',
81         va='top', transform=ax.transAxes)
82  fig.savefig('./figures/stokesOseenFlow.pdf')
83  fig.show()
```

6.9 Three-Dimensional Plots

While matplotlib is primarily a 2D plotting package, it does have basic 3D plotting capabilities. To create a 3D plot, we need to import `Axes3D` from `mpl_toolkits.mplot3d` and then set the keyword `projection` to `'3d'` in a subplot call as follows:

```
import matplotlib.pyplot as plt
from mpl_toolkits.mplot3d import Axes3D
fig = plt.figure()
```

```
ax = fig.add_subplot(111, projection='3d')
```

Different 2D and 3D subplots can be mixed within the same figure window by setting `projection='3d'` only in those subplots where 3D plotting is desired. Alternatively, *all* the subplots in a figure can be set to be 3D plots using the `subplots` function:

```
fig, ax = plt.subplots(subplot_kw={'projection': '3d'})
```

As you might expect, the third axis in a 3D plot is called the *z*-axis, and the same commands for labeling and setting the limits that work for the *x* and *y* axes also work for the *z*-axis.

As a demonstration of matplotlib's 3D plotting capabilities, Fig. 6.22 shows a wireframe and a surface plot of Eq. (6.1), the same equation we plotted with contour plots in Fig. 6.19. The code used to make Fig. 6.22 is given below.

Code: chapter6/programs/wireframeSurfacePlots.py

```
1  import numpy as np
2  import matplotlib.pyplot as plt
3  from mpl_toolkits.mplot3d import Axes3D
4
5
6  def pmgauss(x, y):
7      r1 = (x-1)**2 + (y-2)**2
8      r2 = (x-3)**2 + (y-1)**2
9      return 2*np.exp(-0.5*r1) - 3*np.exp(-2*r2)
10
11
12 a, b = 4, 3
```

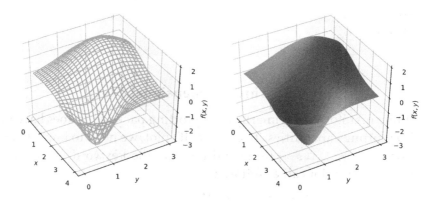

Figure 6.22 Wireframe and surface plots.

```
13
14  x = np.linspace(0, a, 60)
15  y = np.linspace(0, b, 45)
16
17  X, Y = np.meshgrid(x, y)
18  Z = pmgauss(X, Y)
19
20  fig, ax = plt.subplots(1, 2, figsize=(9.2, 4),
21                          subplot_kw={'projection': '3d'})
22  for i in range(2):
23      ax[i].set_zlim(-3, 2)
24      ax[i].xaxis.set_ticks(range(a+1))   # manually set ticks
25      ax[i].yaxis.set_ticks(range(b+1))
26      ax[i].set_xlabel(r'$x$')
27      ax[i].set_ylabel(r'$y$')
28      ax[i].set_zlabel(r'$f(x,y)$')
29      ax[i].view_init(40, -30)
30
31  # Plot wireframe and surface plots.
32  plt.subplots_adjust(left=0.04, bottom=0.04, right=0.96,
33                      top=0.96, wspace=0.05)
34  p0 = ax[0].plot_wireframe(X, Y, Z, rcount=40, ccount=40,
35                            color='C1')
36  p1 = ax[1].plot_surface(X, Y, Z, rcount=50, ccount=50,
37                          color='C1')
38  plt.subplots_adjust(left=0.0)
39  plt.savefig('./figures/wireframeSurfacePlots.pdf')
```

The 3D wireframe and surface plots use the same `meshgrid` function to set up the x-y 2D arrays. The `rcount` and `ccount` keywords set the maximum number of rows and columns used to sample the input data to generate the graph.

These plotting examples are just a sample of many kinds of plots that can be made by matplotlib. Our purpose is not to exhaustively list all the possibilities here but rather to introduce you the matplotlib package. Online documentation and examples are available for you to explore the full range of possibilities.

6.10 Exercises

1. Plot the function $y = 3x^2$ for $-1 \leq x \leq 3$ as a continuous line. Include enough points so that the curve you plot appears smooth. Label the axes x and y.

2. Plot the following function for $-15 \leq x \leq 15$:

$$y = \frac{\cos x}{1 + \frac{1}{5}x^2}$$

   Include enough points so that the curve you plot appears smooth. Draw thin gray lines, one horizontal at $y = 0$ and the other vertical at $x = 0$. Both lines should appear behind the function. Label the axes x and y.

3. Plot the functions $\sin x$ and $\cos x$ vs x on the same plot with x going from $-\pi$ to π. Make sure the limits of the x-axis do not extend beyond the limits of the data. Plot $\sin x$ in the color orange and $\cos x$ in the color green and include a legend to label the two curves. Place the legend within the plot, but such that it does not cover either of the sine or cosine traces. Draw thin gray lines behind the curves, one horizontal at $y = 0$ and the other vertical at $x = 0$.

4. Create a data file with the data shown below.

    (a) Read the data into the Python program and plot t vs. y using circles for data points with error bars. Use the data in the `dy` column as the error estimates for the y data. Label the horizontal and vertical axes "time (s)" and "position (cm)." Create your plot using the `fig, ax = plt.subplots()` syntax.

    (b) On the same graph, plot the function below as a smooth line. Make the line pass *behind* the data points.

    $$y(t) = \left(3 + \frac{1}{2}\sin\frac{\pi t}{5}\right) te^{-t/10}$$

    ```
    Data for Exercise 4
    Date: 16-Aug-2013
    Data taken by Lauren and John

          t         y        dy
        1.0       2.94       0.7
        4.5       8.29       1.2
        8.0       9.36       1.2
    ```

```
                11.5       11.60      1.4
                15.0        9.32      1.3
                18.5        7.75      1.1
                22.0        8.06      1.2
                25.5        5.60      1.0
                29.0        4.50      0.8
                32.5        4.01      0.8
                36.0        2.62      0.7
                39.5        1.70      0.6
                43.0        2.03      0.6
```

5. Use matplotlib's function `hist` along with NumPy's functions `random.rand` and `random.randn` to create the histogram graphs shown in Fig. 9.2. See §9.2 for a description of NumPy's random number functions.

6. The data file below shows data obtained for the displacement (position) *vs.* time of a falling object, together with the estimated uncertainty in the displacement.

```
Measurements of fall velocity vs time
Taken by A.P. Crawford and S.M. Torres
19-Sep-13
time (s)        position (m)        uncertainty (m)
0.0                 0.0                 0.04
0.5                 1.3                 0.12
1.0                 5.1                 0.2
1.5                10.9                 0.3
2.0                18.9                 0.4
2.5                28.7                 0.4
3.0                40.3                 0.5
3.5                53.1                 0.6
4.0                67.5                 0.6
4.5                82.3                 0.6
5.0                97.6                 0.7
5.5               113.8                 0.7
6.0               131.2                 0.7
6.5               148.5                 0.7
7.0               166.2                 0.7
7.5               184.2                 0.7
8.0               201.6                 0.7
8.5               220.1                 0.7
9.0               238.3                 0.7
9.5               256.5                 0.7
10.0              275.6                 0.8
```

(a) Use these data to calculate the velocity and acceleration (in a Python program .py file), together with their uncertainties propagated from the displacement *vs.* time uncertainties. Be sure to calculate time arrays corresponding the midpoint in time between the two displacements or velocities for the velocity and acceleration arrays, respectively.

(b) In a single window frame, make three vertically stacked plots of the displacement, velocity, and acceleration *vs.* time. Show the error bars on the different plots. Make sure that the time axes of all three plots cover the same range of times (use sharex). Why do the relative sizes of the error bars grow progressively greater as one progresses from displacement to velocity to acceleration?

7. Starting from the code that produced Fig. 6.9, write a program using the mixed-OOP syntax introduced in §6.4.1 to produce the plot below. To create this plot, you will need to use the sharex feature introduced in §6.4.1, the subplots_adjust function to adjust the space between the two subplots, and the LATEX syntax introduced in §6.6 to produce the math and Greek symbols. To shorten your program, try to use for loops where there is repetitive code.

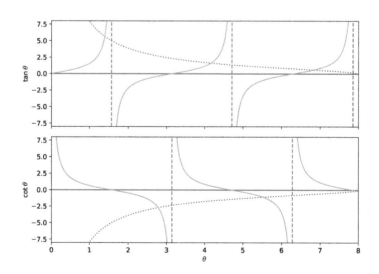

CHAPTER 7

Functions

> In this chapter you learn how to write your own **functions**, similar to the functions provided by Python and NumPy. You learn how to write functions that **process NumPy arrays** efficiently. You learn how to write functions with variable numbers of arguments and how to pass function names (and the arguments of those functions) as an argument of a function you write. We introduce the concept of **namespace**, which isolates the names of variables and functions created inside a function from those created outside the function, with particular attention given to the subtle subject of passing mutable and immutable objects. You learn about anonymous functions (**lambda functions** in Python) and their uses. We continue our discussion of objects and their associated methods and instance variables, here in the context of NumPy arrays. Finally, we illustrate some features of Python functions with their application to least squares fitting.

As you develop more complex computer code, it becomes increasingly important to organize your code into modular blocks. One important means for doing so is *user-defined* Python functions. User-defined functions are a lot like built-in functions that we have encountered in core Python as well as in NumPy and matplotlib. The main difference is that user-defined functions are written by you. The idea is to define functions to simplify your code, improve its readability, and to allow you to reuse the same code in different contexts.

The number of ways that functions are used in programming is so varied that we cannot possibly enumerate all the possibilities. As our use of Python functions in scientific programming is somewhat specialized, we introduce only a few of the possible uses of Python functions, ones that are the most common in scientific programming.

7.1 User-Defined Functions

The NumPy package contains a vast number of mathematical functions. You can find a listing of them at http://docs.scipy.org/doc/numpy/reference/routines.math.html. While the list may seem pretty exhaustive, you may nevertheless find that you need a function that is not available in the NumPy Python library. In those cases, you will want to write your own function.

In studies of optics and signal processing, one often runs into the sinc function, which is defined as

$$\mathrm{sinc}\, x \equiv \frac{\sin x}{x}.$$

Let's write a Python function for the sinc function. Here is our first attempt:

Code: chapter7/programs/sinc0.py

```
import numpy as np
def sinc(x):
    y = np.sin(x)/x
    return y
```

Every function definition begins with the word `def` followed by the name you want to give to the function, `sinc` in this case, then a list of arguments enclosed in parentheses, and finally terminated with a colon. In this case there is only one argument, x, but in general there can be as many arguments as you want, including no arguments at all. For the moment, we will consider the case of just a single argument.

The indented block of code following `def sinc(x):` defines what the function does. In this case, the first line calculates $\mathrm{sinc}\, x = \sin x/x$ and sets it equal to y. The `return` statement of the last line tells Python to return the value of y to the user.

We can try it out in the IPython shell. You can either run the program above that you wrote into a python file or you can type it in—it's only three lines long—into the IPython shell:

```
In [1]: def sinc(x):
   ...:     y = np.sin(x)/x
   ...:     return y
```

We assume you have already imported NumPy. Now the function $\mathrm{sinc}\, x$ is available to be used from the IPython shell.

```
In [2]: sinc(4)
```

Functions

```
Out[2]: -0.18920062382698205

In [3]: a = sinc(1.2)

In [4]: a
Out[4]: 0.77669923830602194

In [5]: np.sin(1.2)/1.2
Out[5]: 0.77669923830602194
```

Inputs and outputs 4 and 5 verify that the function does indeed give the same result as an explicit calculation of sin x/x.

You may have noticed that there is a problem with our definition of sinc x when $x = 0$. Let's try it out and see what happens

```
In [6]: sinc(0.0)
Out[6]: nan
```

IPython returns nan or "not a number," which occurs when Python attempts a division by zero. This is not the desired response as sinc x is, in fact, perfectly well defined for $x = 0$. You can verify this using L'Hopital's rule, which you may have learned in your study of calculus, or you can ascertain the correct answer by calculating the Taylor series for sinc x. Here is what we get:

$$\text{sinc } x = \frac{\sin x}{x} = \frac{x - \frac{x^3}{3!} + \frac{x^5}{5!} + \ldots}{x} = 1 - \frac{x^2}{3!} + \frac{x^4}{5!} + \ldots.$$

From the Taylor series, it is clear that sinc x is well-defined at and near $x = 0$ and that, in fact, sinc(0) = 1. Let's modify our function so that it gives the correct value for x=0.

```
In [7]:  def sinc(x):
   ...:      if x == 0.0:
   ...:          y = 1.0
   ...:      else:
   ...:          y = np.sin(x)/x
   ...:      return y

In [8]: sinc(0)
Out[8]: 1.0

In [9]: sinc(1.2)
Out[9]: 0.77669923830602194
```

Now our function gives the correct value for $x = 0$ as well as for values different from zero.

7.1.1 Looping over arrays in user-defined functions

The code for sinc x works just fine when the argument is a single number or a variable that represents a single number. However, if the argument is a NumPy array, we run into a problem, as illustrated below.

```
In [10]: x = np.arange(0, 5., 0.5)

In [11]: x
Out[11]: array([ 0. , 0.5, 1. , 1.5, 2. , 2.5, 3. ,
        3.5, 4. , 4.5])
In [12]: sinc(x)
Traceback (most recent call last):

  File "<ipython-input-4-b1a03b10f8ff>", line 1, in <module>
    sinc(x)

  File "<ipython-input-1-c944de847889>", line 2, in sinc
    if x==0.0:

ValueError: The truth value of an array with more than one
element is ambiguous. Use a.any() or a.all()
```

The `if` statement in Python is set up to evaluate the truth value of a single variable, not of multi-element arrays. When Python is asked to evaluate the truth value for a multi-element array, it doesn't know what to do and therefore returns an error.

An obvious way to handle this problem is to write the code so that it processes the array one element at a time, which you could do using a `for` loop, as illustrated below.

Code: chapter7/programs/sinc2.py

```
 1  def sinc(x):
 2      y = []                  # empty list to store results
 3      for xx in x:            # loops over in x array
 4          if xx == 0.0:       # appends result of 1.0 to
 5              y += [1.0]      # y list if xx is zero
 6          else:               # appends result of sin(xx)/xx to y
 7              y += [np.sin(xx)/xx]  # list if xx is not zero
 8      return np.array(y)      # converts y to array and
 9
10
11      # returns array
12  import numpy as np
13  import matplotlib.pyplot as plt
14
15  x = np.linspace(-10, 10, 255)
16  y = sinc(x)
```

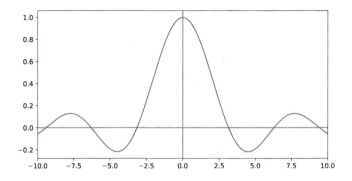

Figure 7.1   Plot of user-defined sinc(x) function.

```
17
18  fig, ax = plt.subplots(figsize=(8, 4))
19  ax.plot(x, y)
20  ax.set_xlim(-10, 10)
21  ax.axhline(color="gray", zorder=-1)
22  ax.axvline(color="gray", zorder=-1)
23  fig.savefig("sinc2.pdf")
24  fig.show()
```

The for loop evaluates the elements of the x array one by one and appends the results to the list y one by one. When it is finished, it converts the list to an array and returns the array. The code following the function definition plots sinc x as a function of x.

In the program above, you may have noticed that the NumPy library is imported *after* the sinc(x) function definition. As the function uses the NumPy functions sin and array, you may wonder how this program can work. Doesn't the import numpy statement have to be called before any NumPy functions are used? The answer is an emphatic "YES." What you need to understand is that the function definition is *not executed* when it is defined, nor can it be, as it has no input x data to process. That part of the code is just a definition. The first time the code for the sinc(x) function is actually executed is when it is called on line 16 of the program, which occurs after the NumPy library is imported in line 12. Figure 7.1 shows the plot of the sinc(x) function generated by the above code.

7.1.2 Fast array processing for user-defined functions

While using loops to process arrays works just fine, it is usually not the best way to accomplish a task in Python. The reason is that loops in Python are executed rather slowly, as we saw in §5.2.3. To deal with this problem, the developers of NumPy introduced a number of functions designed to process arrays quickly and efficiently. For the present case, what we need is a conditional statement or function that can process arrays directly. The function we want is called `where` and it is a part of the NumPy library. The `where` function has the form

```
where(condition, output if True, output if False)
```

The first argument of the `where` function is a conditional statement involving an array. The `where` function applies the condition to the array element by element, and returns the second argument for those array elements for which the condition is `True`, and returns the third argument for those array elements that are `False`. We can apply it to the `sinc(x)` function as follows

Code: chapter7/programs/sincTest.py

```
 1  import numpy as np
 2  import matplotlib.pyplot as plt
 3
 4
 5  def sinc(x):
 6      z = np.where(x == 0.0, 1.0, np.sin(x)/x)
 7      return z
 8
 9
10  x = np.linspace(-10, 10, 255)
11  y = sinc(x)
12
13  fig, ax = plt.subplots(figsize=(8, 4))
14  ax.plot(x, y)
15  ax.axhline(color="gray", zorder=-1)
16  ax.axvline(color="gray", zorder=-1)
17  fig.show()
```

The `where` function creates the array `y` and sets the elements of `y` equal to 1.0 where the corresponding elements of x are zero, and otherwise sets the corresponding elements to `sin(x)/x`. This code executes much faster, 25 to 100 times or more, depending on the size of the array, than the code using a `for` loop. Moreover, the new code is much simpler to write and read. An additional benefit of the `where` function is that it can handle single variables and arrays equally well. The code we wrote for the sinc function with the `for` loop cannot handle single variables. Of course we could rewrite the code so that it did, but the code becomes even more clunky. It's better just to use NumPy's `where` function.

The moral of the story

The moral of the story is that you should avoid using `for` and `while` loops to process arrays in Python programs whenever an array-processing method is available. As a beginning Python programmer, you may not always see how to avoid loops, and indeed, avoiding them is not always possible. But you should look for ways to avoid them, especially loops that iterate a large number of times. As you become more experienced, you will find that using array-processing methods in Python becomes more natural. Using them can greatly speed up the execution of your code, especially when working with large arrays.

Vectorized code and ufuncs

Finally, a word about jargon. Programmers sometimes refer to using array-processing methods as *vectorizing* code. The jargon comes from the idea that an array of N elements can be regarded as an N-dimensional vector. Computer code that processes *vectors* as the basic unit rather than individual data elements is said to be *vectorized*.

NumPy functions are always vectorized and are known as *universal functions* or *ufuncs* for short.

Don't worry too much about the jargon or even its origin. But it's useful to understand when reading from different sources, online or otherwise, about Python code.

7.1.3 Functions with more than one input or output

Python functions can have any number of input arguments and can return any number of variables. For example, suppose you want a function that outputs (x, y)-coordinates of n points evenly distributed around a circle of radius r centered at the point (x_0, y_0). The inputs to the function would be r, x_0, y_0, and n. The outputs would be the (x, y)-coordinates. The following code implements this function.

Code: chapter7/programs/circleN.py

```
import numpy as np
```

```
 2
 3
 4  def circle(r, x0, y0, n):
 5      theta = np.linspace(0., 2*np.pi, n, endpoint=False)
 6      x, y = r * np.cos(theta), r * np.sin(theta)
 7      return x0+x, y0+y
```

This function has four inputs and two outputs. In this case, the four inputs are simple numeric variables and the two outputs are NumPy arrays. In general, the inputs and outputs can be any combination of data types: arrays, lists, strings, *etc*. Of course, the body of the function must be written to be consistent with the prescribed data types.

7.1.4 Positional and keyword arguments

It is often useful to have function arguments that have some default setting. This happens when you want an input to a function to have some standard value or setting most of the time, but you would like to reserve the possibility of giving it some value other than the default value.

For example, in the program `circle` from the previous section, we might decide that under most circumstances, we want n=12 points around the circle, like the points on a clock face, and we want the circle to be centered at the origin. In this case, we would rewrite the code to read

Code: chapter7/programs/circleKW.py

```
 1  import numpy as np
 2
 3
 4  def circle(r, x0=0.0, y0=0.0, n=12):
 5      theta = np.linspace(0., 2*np.pi, n, endpoint=False)
 6      x, y = r * np.cos(theta), r * np.sin(theta)
 7      return x0+x, y0+y
```

The default values of the arguments `x0`, `y0`, and `n` are specified in the argument of the function definition in the `def` line. Arguments whose default values are specified in this manner are called *keyword arguments*, and they can be omitted from the function call if the user is content using those values. For example, writing `circle(4)` is now a perfectly legal way to call the `circle` function and it would produce 12 (x,y) coordinates centered about the origin $(x,y) = (0,0)$. On the other hand, if you want the values of `x0`, `y0`, and `n` to be something

Functions

different from the default values, you can specify their values as you would have before.

If you want to change only some of the keyword arguments, you can do so by using the keywords in the function call. For example, suppose you are content with having the circle centered on $(x, y) = (0, 0)$ but you want only 6 points around the circle rather than 12. Then you would call the `circle` function as follows:

```
circle(2, n=6)
```

The unspecified keyword arguments keep their default values of zero but the number of points `n` around the circle is now 6 instead of the default value of 12.

The normal arguments without keywords are called *positional arguments*; they have to appear *before* any keyword arguments and, when the function is called, must appear in the same order as specified in the function definition. The keyword arguments, if supplied, can appear in any order provided they appear with their keywords. If supplied without their keywords, then they must also appear in the order they appear in the function definition. The following function calls to `circle` both give the same output.

```
In [13]: circle(3, n=3, y0=4, x0=-2)
Out[13]: (array([ 1. ,  -3.5,   -3.5]),
         array([ 4. ,  6.59807621, 1.40192379]))

In [14]: circle(3, -2, 4, 3)    # w/o keywords, arguments
                                # supplied in order
Out[14]: (array([ 1. ,  -3.5,   -3.5]),
         array([ 4. ,  6.59807621, 1.40192379]))
```

By now you probably have noticed that we used the keyword argument `endpoint` in calling `linspace` in our definition of the `circle` function. The default value of `endpoint` is `True`, meaning that `linspace` includes the endpoint specified in the second argument of `linspace`. We set it equal to `False` so that the last point was not included. Do you see why?

7.1.5 Variable number of arguments

While it may seem odd, it is sometimes useful to leave the number of arguments unspecified. A simple example is a function that computes the product of an arbitrary number of numbers:

```
def product(*args):
```

```
print("args = {}".format(args))
p = 1
for num in args:
    p *= num
return p
```

Placing the "*" before the args argument tells Python that args is an unnamed argument that can have any number of entries. For example, here we give it three entries:

```
In [15]: product(11., -2, 3)
args = (11.0, -2, 3)
Out[15]:   -66.0
```

Here we give it only two arguments:

```
In [16]: product(2.31, 7)
args = (2.31, 7)
Out[16]:    16.17
```

The print("args...) statement in the function definition is not necessary, of course, but is put in to show that the argument args is a tuple inside the function. Here, the *args tuple argument is used because one does not know ahead of time how many numbers are to be multiplied together.

7.1.6 Passing function names and parameters as arguments

The *args tuple argument is also quite useful in another context: when passing the name of a function as an argument in another function. In many cases, the function name that is passed may have a number of parameters that must also be passed but aren't known ahead of time. If this all sounds a bit confusing—functions calling other functions with *arbitrary* parameters—a concrete example will help you understand.

Suppose we have the following function that numerically computes the value of the derivative of an arbitrary function $f(x)$:

Code: chapter7/programs/derivA.py

```
1  def deriv(f, x, h=1.e-9, *params):
2      return (f(x + h, *params) - f(x - h, *params))/(2.*h)
```

The argument *params is an optional *positional* argument. We begin by demonstrating the use of the function deriv without using the optional *params argument. Suppose we want to compute the derivative of the function $f_0(x) = 4x^5$. First, we define the function:

```
In [17]: def f0(x):
    ...:     return 4*x**5
```

Now let's find the derivative of $f_0(x) = 4x^5$ at $x = 3$ using the function `deriv`:

```
In [18]: deriv(f0, 3)
Out[18]: 1620.0001482502557
```

The exact result is 1620, so our function to numerically calculate the derivative works pretty well (it's accurate to about 1 part in 10^7).

Suppose we had defined a more general function $f_1(x) = ax^p$ as follows:

```
In [19]: def f1(x, a, p):
    ...:     return a*x**p
```

Suppose we want to calculate the derivative of this function for a particular set of numerical values of the parameters a and p. Now we face a problem, because it might seem that there is no way to pass the values of the parameters a and p to the `deriv` function. Moreover, this is a generic problem for functions such as `deriv` that use a function as an input, because different functions you might want to use as inputs generally come with a different number of parameters. Therefore, we would like to write our program `deriv` so that it works, irrespective of how many parameters are needed to specify a particular function.

This is what the optional positional argument `*params` defined in `deriv` is for: to pass parameters of `f1`, like a and p, through `deriv`. To see how this works, let's set a and p to be 4 and 5, respectively, the same values we used in the definition of `f0`, so that we can compare the results:

```
In [20]: deriv(f1, 3, 1.e-9, *(4, 5))
Out[20]: 1620.0001482502557
```

We get the same answer as before, but this time we have used `deriv` with a more general form of the function $f_1(x) = ax^p$.

The order of the parameters a and p is important. The function `deriv` uses x, the first argument of `f1`, as its principal argument, and then uses a and p, in the same order that they are defined in the function `f1`, to fill in the additional arguments—the parameters—of the function `f1`.

Beware, the `params` argument *must* be a tuple. If there is only one parameter, as there is for the function $g(x) = (x + a)/(x - a)$, then the call to the derivative function would work like this:

```
In [21]: def g(x, a):
   ...:     return (x+a)/(x-a)

In [22]: a = 1.0

In [23]: x = np.linspace(0, 2, 6)

In [24]: deriv(g, x, 1.e-9, *(a, ))
Out[24]:
array([ -2.00000011,  -5.55555557, -49.99999792, -50.00000414,
        -5.55555602,  -2.00000017])
```

The comma following a in the argument *(a,) is needed so that (a,) is understood by Python to be a tuple.

Optional arguments must appear after the regular positional and keyword arguments in a function call. The order of the arguments must adhere to the following convention:

```
def func(pos1, pos2, ..., keywd1, keywd2,
        ..., *args, **kwargs):
```

That is, the order of arguments is: positional arguments first, then keyword arguments, then optional positional arguments (*args), then optional keyword arguments (**kwargs). Note that to use the *params argument, we had to explicitly include the keyword argument h even though we didn't need to change it from its default value.

Python also allows for a variable number of keyword arguments—**kwargs—in a function call, that is, an argument preceded by **. While args is a tuple, kwargs is a dictionary, so the value of an optional keyword argument is accessed through its dictionary key. To use the **kwargs format, we rewrite our deriv function using two stars (**params):

Code: chapter7/programs/derivK.py

```
def deriv(f, x, h=1.e-9, **params):
    return (f(x + h, **params) - f(x - h, **params))/(2.*h)
```

Next we define a dictionary:

```
In [25]: d = {'a': 4, 'p': 5}
```

And then finally, we input our optional keyword arguments using **d:

```
In [26]: deriv(f1, 3, **d)
Out[26]: 1620.0001482502557
```

We can also include our optional keyword arguments as a dictionary literal:

Functions

```
In [27]: deriv(f1, 3, **{'a': 4, 'p': 5})
Out[27]:1620.0001482502557
```

Note that when using `**kwargs`, you can omit keyword arguments, in this case `h`, if you want to use the default value(s).

7.2 Passing data (objects) to and from functions

Functions are like mini-programs within the larger programs that call them. Each function has a set of variables with certain names that are to some degree or other isolated from the calling program. We shall get more specific about just how isolated those variables are below, but before we do, we introduce the concept of a *namespace*. Each function has its own namespace, which is essentially a mapping of variable names to objects, like numerics, strings, lists, and so forth. It's a kind of dictionary. The calling program has its own namespace, distinct from that of any functions it calls. The distinctiveness of these namespaces plays an important role in how functions work, as we shall see below.

7.2.1 Variables and arrays created entirely within a function

An important feature of functions is that variables and arrays created *entirely within* a function cannot be seen by the program that calls the function unless the variable or array is explicitly passed to the calling program in the `return` statement. This is important because it means you can create and manipulate variables and arrays, giving them any name you please, without affecting any variables or arrays outside the function, even if the variables and arrays inside and outside a function share the same name.

To see what how this works, let's rewrite our program to plot the `sinc` function using the `sinc` function definition that uses the `where` function.

Code: chapter7/programs/sincTest.py

```
1  import numpy as np
2  import matplotlib.pyplot as plt
3
4
5  def sinc(x):
6      z = np.where(x == 0.0, 1.0, np.sin(x)/x)
7      return z
```

```
 8
 9
10  x = np.linspace(-10, 10, 255)
11  y = sinc(x)
12
13  fig, ax = plt.subplots(figsize=(8, 4))
14  ax.plot(x, y)
15  ax.axhline(color="gray", zorder=-1)
16  ax.axvline(color="gray", zorder=-1)
17  fig.show()
```

We save this program in a file named sincTest.py. Running this program, by typing run sincTest.py in the IPython terminal, produces a plot like the plot of sinc shown in Fig. 7.1. Notice that the array variable z is only defined within the function definition of sinc. Running the program from the IPython terminal produces the plot, of course. Then if we ask IPython to print out the arrays, x, y, and z, we get some interesting and informative results, as shown below.

```
In [1]: run sincTest.py

In [2]: x
Out[2]: array([-10. , -9.99969482, -9.99938964,
        ..., 9.9993864, 9.99969482, 10. ])

In [3]: y
Out[3]: array([-0.05440211, -0.05437816, -0.0543542 ,
        ..., -0.0543542 , -0.05437816, -0.05440211])

In [4]: z
-------------------------------------------------
NameError       Traceback (most recent call last)

NameError: name 'z' is not defined
```

When we type in x at the In [2]: prompt, IPython prints out the array x (some of the output is suppressed because the array x has many elements); similarly for y. But when we type z at the In [4]: prompt, IPython returns a NameError because z is not defined. The IPython terminal is working in the same *namespace* as the program. But the namespace of the sinc function is isolated from the namespace of the program that calls it, and therefore isolated from IPython. This also means that when the sinc function ends with return z, it doesn't return the name z, but instead assigns the values in the array z to the array y, as directed by the main program in line 11.

7.2.2 Passing lists and arrays to functions: Mutable and immutable objects

What happens to a variable or an array passed to a function when the variable or array is *changed* within the function? It turns out that the answers are different depending on whether the variable passed is a simple numeric variable, string, or tuple, or whether it is an array or list. The program below illustrates the different ways that Python handles single variables *vs* the way it handles lists and arrays.

Code: chapter7/programs/passingVars.py

```
 1  def test(s, v, t, l, a):
 2      s = "I am doing fine"
 3      v = np.pi**2
 4      t = (1.1, 2.9)
 5      l[-1] = 'end'
 6      a[0] = 963.2
 7      return s, v, t, l, a
 8
 9
10  import numpy as np
11
12  s = "How do you do?"
13  v = 5.0
14  t = (97.5, 82.9, 66.7)
15  l = [3.9, 5.7, 7.5, 9.3]
16  a = np.array(l)
17
18  print('**************')
19  print("s = {0:s}".format(s))
20  print("v = {0:5.2f}".format(v))
21  print("t = {}".format(t))
22  print("l = {}".format(l))
23  print("a = "),              # comma suppresses line feed
24  print(a)
25  print('**************')
26  print('*call "test"*')
27
28  s1, v1, t1, l1, a1 = test(s, v, t, l, a)
29
30  print('**************')
31  print("s1 = {0:s}".format(s1))
32  print("v1 = {0:5.2f}".format(v1))
33  print("t1 = {}".format(t1))
34  print("l1 = {}".format(l1))
35  print("a1 = "),             # comma suppresses line feed
36  print(a1)
37  print('**************')
```

```
38  print("s = {0:s}".format(s))
39  print("v = {0:5.2f}".format(v))
40  print("t = {}".format(t))
41  print("l = {}".format(l))
42  print("a = "),
43  print(a)
44  print('*************')
```

The function `test` has five arguments, a string `s`, a numerical variable `v`, a tuple `t`, a list `l`, and a NumPy array `a`. `test` modifies each of these arguments and then returns the modified `s`, `v`, `t`, `l`, `a`. Running the program produces the following output.

```
In [17]: run passingVars.py


*************
s = How do you do?
v =  5.00
t = (97.5, 82.9, 66.7)
l = [3.9, 5.7, 7.5, 9.3]
a =
[3.9 5.7 7.5 9.3]
*************
*call "test"*
*************
s1 = I am doing fine
v1 =  9.87
t1 = (1.1, 2.9)
l1 = [3.9, 5.7, 7.5, 'end']
a1 =
[963.2   5.7   7.5   9.3]
*************
s = How do you do?
v =  5.00
t = (97.5, 82.9, 66.7)
l = [3.9, 5.7, 7.5, 'end']
a =
[963.2   5.7   7.5   9.3]
*************
```

The program prints out three blocks of variables separated by asterisks. The first block merely verifies that the contents of `s`, `v`, `t`, `l`, and `a` are those assigned in lines 12–16. Then the function `test` is called. The next block prints the output of the call to the function `test`, namely the variables `s1`, `v1`, `t1`, `l1`, and `a1`. The results verify that the function modified the inputs as directed by the `test` function.

The third block prints out the variables `s`, `v`, `t`, `l`, and `a` from the calling program *after* the function `test` was called. These variables served as the inputs to the function `test`. Examining the output from the third printing block, we see that the values of the string `s`, the numeric variable `v`, and the contents of `t` are unchanged after the function call. This is probably what you would expect. On the other hand, we see that the list `l` and the array `a` are changed after the function call. This might surprise you! But these are important points to remember, so it is important that we summarize them in two bullet points here:

- Changes to string, variable, and tuple arguments of a function within the function do not affect their values in the calling program.

- Changes to values of elements in list and array arguments of a function within the function are reflected in the values of the same list and array elements in the calling function.

The point is that simple numerics, strings and tuples are immutable while lists and arrays are mutable. Because immutable objects can't be changed, changing them within a function creates new objects with the same name inside of the function, but the old immutable objects that were used as arguments in the function call remain unchanged in the calling program. On the other hand, if elements of mutable objects like those in lists or arrays are changed, then those elements that are changed inside the function are also changed in the calling program.

7.3 Anonymous Functions: `lambda` Expressions

Python provides another way to generate functions called *lambda* expressions. A lambda expression is a kind of in-line function that can be generated on the fly to accomplish some small task, often where a function name is needed as input to another function and thus is used only once.

You can assign a lambda expression a name, but you don't need to; hence, they are sometimes called *anonymous* functions.

A lambda expression uses the keyword `lambda` and has the general form

```
lambda arg1, arg2, ... : output
```

The arguments arg1, arg2, ... are inputs to a lambda, just as for a functions, and the output is an expression using the arguments.

While lambda expressions need not be named, we illustrate their use by comparing a conventional Python function definition to a lambda expression to which we give a name. First, we define a conventional Python function:

```
In [1]: def f(a, b):
   ...:     return 3*a + b**2

In [2]: f(2, 3)
Out[2]: 15
```

Next, we define a lambda expression that does the same thing:

```
In [3]: g = lambda a, b: 3*a + b**2

In [4]: g(2, 3)
Out[4]: 15
```

The lambda expression defined by g does the same thing as the function f. Such lambda expressions are useful when you need a very short function definition, usually to be used locally only once or perhaps a few times.

Lambda expressions can be useful as function arguments, particularly when extra parameters need to be passed with the function. In §7.1.6, we saw how Python functions can do this using optional arguments, *args and **kwargs. Lambda expressions provide another means for accomplishing the same thing. To see how this works, recall our definition of the function to take the derivative of another function:

Code: chapter7/programs/derivA.py

```
1  def deriv(f, x, h=1.e-9, *params):
2      return (f(x + h, *params) - f(x - h, *params))/(2.*h)
```

and our definition of the function

```
In [5]: def f1(x, a, p):
   ...:     return a*x**p
```

Instead of using the *params optional argument to pass the values of the parameters a and p, we can define a lambda expression that is a function of x alone, with a and p set in the lambda expression.

```
In [6]: g = lambda x: f1(x, 4, 5)
```

The function g defined by the lambda expression is the same as f1(x, a, p) but with a and p set to 4 and 5, respectively. Now we can use deriv to calculate the derivative at $x = 3$ using the lambda function g

```
In [7]: deriv(g, 3)
Out[7]: 1620.0001482502557
```

Of course, we get the same answer as we did using the other methods.

You might wonder why we can't just insert f1(x, 4, 5) as the argument to deriv. The reason is that you need to pass the *name* of the function, not the function itself. We assign the name g to our lambda expression, and then pass that name through the argument of deriv.

Alternatively, we can simply insert the whole lambda expression in the argument of deriv where the function name goes:

```
In [8]: deriv(lambda x: f1(x, 4, 5), 3)
Out[8]: 1620.0001482502557
```

This works too. In this case, however, we never defined a function name for our lambda expression. Our lambda expression is indeed an *anonymous function*.

You may recall that we already used lambda expressions in §4.2.2 where we discussed how to print formatted arrays. There are also a number of nifty programming tricks that can be implemented using lambda expressions, but we will not go into them here. Look up lambdas on the web if you are curious about their more exotic uses.

7.4 NumPy Object Attributes: Methods and Instance Variables

You have already encountered quite a number of functions that are part of either NumPy or Python or matplotlib. But there is another way in which Python implements things that act like functions: these are the *methods* associated with an object that we introduced in §3.5. Recall from §3.5 that strings, arrays, lists, and other such data structures in Python are not merely the numbers or strings we have defined them to be. They are *objects*. In general, an object in Python has associated with it a number of *attributes*, which are either *instance variables* associated with the object or specialized functions called *methods* that act on the object.

Let's start with the NumPy array. A NumPy array is a Python object and therefore has associated with it a number of attributes: instance variables and methods. Suppose, for example, we write a

= random.random(10), which creates an array of 10 uniformly distributed random numbers between 0 and 1.[1] An example of an instance variable associated with an array is the size or number of elements in the array. An instance variable of an object in Python is accessed by typing the object name followed by a period followed by the variable name. The code below illustrates how to access two different instance variables of an array, its size and its data type.

```
In [1]: a = random.random(10)

In [2]: a.size
Out[2]: 10

In [3]: a.dtype
Out[3]: dtype('float64')
```

Any object in Python can, and in general does, have a number of instance variables that are accessed in just the way demonstrated above, with a period and the instance variable name following the name of the particular object. In general, instance variables involve properties of the object that are stored by Python with the object and require no computation. Python just looks up the attribute and returns its value.

Objects in Python also have associated with them a number of specialized functions called *methods* that act on the object. Methods generally involve Python performing some kind of computation. Methods are accessed in a fashion similar to instance variables, by appending a period followed the method's name, which is followed by a pair of open-close parentheses, consistent with methods being a function that acts on the object. Often methods are used with no arguments, as methods by default act on the object whose name they follow. In some cases, however, methods can take arguments. Examples of methods for NumPy arrays are sorting, calculating the mean, or standard deviation of the array. The code below illustrates a few array methods.

```
In [21]: a
Out[21]:
array([ 0.859057 , 0.27228037, 0.87780026,
        0.14341207, 0.05067356, 0.83490135,
        0.54844515, 0.33583966, 0.31527767,
        0.15868803])

In [22]: a.sum()                        # sum
```

---

[1] NumPy's random module has a number of routines for generating arrays of random numbers, as discussed in §9.2.

```
Out[22]: 4.3963751104791005

In [23]: a.mean()              # mean or average
Out[23]: 0.43963751104791005

In [24]: a.var()               # variance
Out[24]: 0.090819477333711512

In [25]: a.std()               # standard deviation
Out[25]: 0.30136270063448711

In [26]: a.sort()              # sort small to large

In [27]: a
Out[27]:
array([ 0.05067356, 0.14341207, 0.15868803,
        0.27228037, 0.31527767, 0.33583966,
        0.54844515, 0.83490135, 0.859057,
        0.87780026])
```

Notice that the `sort()` method has permanently changed the order of the elements of the array.

```
In [28]: a.clip(0.3, 0.8)
Out[29]:
array([ 0.3, 0.3, 0.3, 0.3, 0.31527767, 0.33583966,
        0.54844515, 0.8, 0.8    , 0.8])
```

The `clip()` method provides an example of a method that takes an argument, in this case the arguments are the lower and upper values to which array elements are cut off if their values are outside the range set by these values.

7.5 Example: Linear Least Squares Fitting

In this section we illustrate how to use functions and methods in the context of modeling experimental data.

In science and engineering we often have some theoretical curve or *fitting function* that we would like to fit to some experimental data. In general, the fitting function is of the form $f(x; a, b, c, ...)$, where x is the independent variable and $a, b, c, ...$ are parameters to be adjusted so that the function $f(x; a, b, c, ...)$ best fits the experimental data. For example, suppose we had some data of the velocity *vs* time for a falling mass. If the mass falls only a short distance, such that its velocity remains well below its terminal velocity, we can ignore air resistance. In

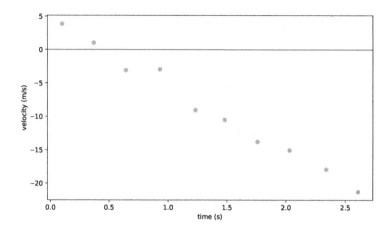

Figure 7.2   Velocity *vs* time for falling mass.

this case, we expect the acceleration to be constant and the velocity to change linearly in time according to the equation

$$v(t) = v_0 - gt, \tag{7.1}$$

where g is the local gravitational acceleration. We can fit the data graphically, say by plotting it as shown in Fig. 7.2, and then drawing a line through the data. When we draw a straight line through the data, we try to minimize the distance between the points and the line, globally averaged over the whole data set.

While this can give a reasonable estimate of the best fit to the data, the procedure is rather *ad hoc*. We would prefer to have a more well-defined analytical method for determining what constitutes a "best fit." One way to do that is to consider the sum

$$S = \sum_i^n [y_i - f(x_i; a, b, c, ...)]^2, \tag{7.2}$$

where y_i and $f(x_i; a, b, c, ...)$ are the values of the experimental data and the fitting function, respectively, at x_i, and S is the square of their difference summed over all n data points. The quantity S is a sort of global measure of how much the fit $f(x_i; a, b, c, ...)$ differs from the experimental data y_i.

Notice that for a given set of data points $\{x_i, y_i\}$, S is a function

Functions

only of the fitting parameters $a, b, ...$, that is, $S = S(a, b, c, ...)$. One way of defining a *best* fit, then, is to find the set of values of the fitting parameters $a, b, ...$ that minimize the value of S.

In principle, finding the values of the fitting parameters $a, b, ...$ that minimize the S is a simple matter. Just set the partial derivatives of S with respect to the fitting parameter equal to zero and solve the resulting system of equations:

$$\frac{\partial S}{\partial a} = 0, \quad \frac{\partial S}{\partial b} = 0, ... \quad (7.3)$$

Because there are as many equations as there are fitting parameters, we should be able to solve the system of equations and find the values of the fitting parameters that minimize S. Solving those systems of equations is straightforward if the fitting function $f(x; a, b, ...)$ is linear in the fitting parameters. Some examples of fitting functions linear in the fitting parameters are:

$$f(x; a, b) = a + bx$$
$$f(x; a, b, c) = a + bx + cx^2$$
$$f(x; a, b, c) = a \sin x + be^x + ce^{-x^2}. \quad (7.4)$$

For fitting functions such as these, taking the partial derivatives with respect to the fitting parameters, as proposed in Eq. (7.3), results in a set of algebraic equations that are linear in the fitting parameters $a, b, ...$ Because they are linear, these equations can be solved in a straightforward manner.

For cases in which the fitting function is not linear in the fitting parameters, one can generally still find the values of the fitting parameters that minimize S, but finding them requires more work, which goes beyond our immediate interests here.

7.5.1 Linear regression

We start by considering the simplest case, fitting a straight line to a set of $\{x_i, y_i\}$ data, such as the data set shown in Fig. 7.2. Here the fitting function is $f(x) = a + bx$, which is linear in the fitting parameters a and b. For a straight line, the sum in Eq. (7.2) becomes

$$S(a, b) = \sum_i (y_i - a - bx_i)^2, \quad (7.5)$$

where the sum is over all the points in the $\{x_i, y_i\}$ data set. Finding the best fit in this case corresponds to finding the values of the fitting parameters a and b for which $S(a,b)$ is a minimum. To find the minimum, we set the derivatives of $S(a,b)$ equal to zero:

$$\frac{\partial S}{\partial a} = \sum_i -2(y_i - a - bx_i) \quad = 2(na + b\sum_i x_i - \sum_i y_i) = 0$$
$$\frac{\partial S}{\partial b} = \sum_i -2(y_i - a - bx_i)x_i \quad = 2\left(a\sum_i x_i + b\sum_i x_i^2 - \sum_i x_1 y_i\right) = 0 \qquad (7.6)$$

Dividing both equations by $2n$ leads to the equations

$$a + b\bar{x} = \bar{y}$$
$$a\bar{x} + b\frac{1}{n}\sum_i x_i^2 = \frac{1}{n}\sum_i x_i y_i \qquad (7.7)$$

where

$$\bar{x} = \frac{1}{n}\sum_i x_i, \quad \bar{y} = \frac{1}{n}\sum_i y_i. \qquad (7.8)$$

Solving Eq. (7.7) for the fitting parameters gives

$$b = \frac{\sum_i x_i y_i - n\bar{x}\bar{y}}{\sum_i x_i^2 - n\bar{x}^2}, \quad a = \bar{y} - b\bar{x}. \qquad (7.9)$$

Noting that $n\bar{y} = \sum_i y$ and $n\bar{x} = \sum_i x$, the results can be written as

$$b = \frac{\sum_i (x_i - \bar{x})y_i}{\sum_i (x_i - \bar{x})x_i}, \quad a = \bar{y} - b\bar{x}. \qquad (7.10)$$

While Eqs. (7.9) and (7.10) are equivalent analytically, Eq. (7.10) is preferred for numerical calculations because Eq. (7.10) is less sensitive to roundoff errors. Here is a Python function implementing this algorithm:

Code: chapter7/programs/lineFit.py

```
def lineFit(x, y):
    ''' Returns slope and y-intercept of linear fit to (x,y)
    data set'''
    xavg = x.mean()
    slope = (y * (x-xavg)).sum()/(x * (x-xavg)).sum()
    yint = y.mean() - slope*xavg
    return slope, yint
```

It's hard to imagine a simpler implementation of the linear regression algorithm.

7.5.2 Linear regression with weighting: χ^2

The linear regression routine of the previous section weights all data points equally. That is fine if the absolute uncertainty is the same for all data points. In many cases, however, the uncertainty is different for different points in a data set. In such cases, we would like to weight the data that has smaller uncertainty more heavily than those data that have greater uncertainty. For this case, there is a standard method of weighting and fitting data that is known as χ^2 (or *chi-squared*) fitting. In this method we suppose that associated with each (x_i, y_i) data point is an uncertainty in the value of y_i of $\pm\sigma_i$. In this case, the "best fit" is defined as the one with the set of fitting parameters that minimizes the sum

$$\chi^2 = \sum_i \left(\frac{y_i - f(x_i)}{\sigma_i}\right)^2. \tag{7.11}$$

Setting the uncertainties $\sigma_i = 1$ for all data points yields the same sum S that we introduced in the previous section. In this case, all data points are weighted equally. However, if σ_i varies from point to point, it is clear that those points with large σ_i contribute less to the sum than those with small σ_i. Thus, data points with large σ_i are weighted less than those with small σ_i.

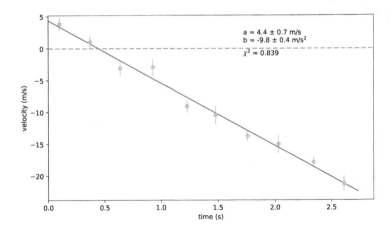

Figure 7.3 Fit using χ^2 least squares fitting routine with data weighted by error bars.

To fit data to a straight line, we set $f(x) = a + bx$ and write

$$\chi^2(a,b) = \sum_i \left(\frac{y_i - a - bx_i}{\sigma_i} \right)^2. \tag{7.12}$$

Finding the minimum for $\chi^2(a,b)$ follows the same procedure used for finding the minimum of $S(a,b)$ in the previous section. The result is

$$b = \frac{\sum_i (x_i - \hat{x}) y_i / \sigma_i^2}{\sum_i (x_i - \hat{x}) x_i / \sigma_i^2}, \quad a = \hat{y} - b\hat{x}. \tag{7.13}$$

where

$$\hat{x} = \frac{\sum_i x_i / \sigma_i^2}{\sum_i 1/\sigma_i^2}, \quad \hat{y} = \frac{\sum_i y_i / \sigma_i^2}{\sum_i 1/\sigma_i^2}. \tag{7.14}$$

For a fit to a straight line, the overall quality of the fit can be measured by the reduced chi-squared parameter

$$\chi_r^2 = \frac{\chi^2}{n-2} \tag{7.15}$$

where χ^2 is given by Eq. (7.11) evaluated at the optimal values of a and b given by Eq. (7.13). A good fit is characterized by $\chi_r^2 \approx 1$. This makes sense because if the uncertainties σ_i have been properly estimated, then $[y_i - f(x_i)]^2$ should on average be roughly equal to σ_i^2, so that the sum in Eq. (7.11) should consist of n terms approximately equal to 1. Of course, if there were only 2 terms ($n = 2$), then χ^2 would be zero as the best straight line fit to two points is a perfect fit. That is essentially why χ_r^2 is normalized using $n - 2$ instead of n. If χ_r^2 is significantly greater than 1, this indicates a poor fit to the fitting function (or an underestimation of the uncertainties σ_i). If χ^2 is significantly less than 1, then it indicates that the uncertainties were probably overestimated (the fit and fitting function may or may not be good).

We can also get estimates of the uncertainties in our determination of the fitting parameters a and b, although deriving the formulas is a bit more involved than we want to get into here. Therefore, we just give the results:

$$\sigma_b^2 = \frac{1}{\sum_i (x_i - \hat{x}) x_i / \sigma_i^2}, \quad \sigma_a^2 = \sigma_b^2 \frac{\sum_i x_i^2 / \sigma_i^2}{\sum_i 1/\sigma_i^2}. \tag{7.16}$$

Functions

The estimates of uncertainties in the fitting parameters depend explicitly on $\{\sigma_i\}$ and will only be meaningful if (*i*) $\chi_r^2 \approx 1$ and (*ii*) the estimates of the uncertainties σ_i are accurate.

You can find more information, including a derivation of Eq. (7.16), in *Data Reduction and Error Analysis for the Physical Sciences, 3rd ed* by P. R. Bevington & D. K. Robinson, McGraw-Hill, New York, 2003.

7.6 Exercises

1. Write a function that can return each of the first three spherical Bessel functions $j_n(x)$:

$$j_0(x) = \frac{\sin x}{x}$$
$$j_1(x) = \frac{\sin x}{x^2} - \frac{\cos x}{x} \qquad (7.17)$$
$$j_2(x) = \left(\frac{3}{x^2} - 1\right)\frac{\sin x}{a} - \frac{3\cos x}{x^2}$$

Your function should take as arguments a NumPy array x and the order n, and should return an array of the designated order n spherical Bessel function. Take care to make sure that your functions behave properly at $x = 0$.

Demonstrate the use of your function by writing a Python routine that plots the three Bessel functions for $0 \le x \le 20$. Your plot should look like the one below. Something to think about: You might note that $j_1(x)$ can be written in terms of $j_0(x)$, and that $j_2(x)$ can be written in terms of $j_1(x)$ and $j_0(x)$. Can you take advantage of this to write a more efficient function for the calculations of $j_1(x)$ and $j_2(x)$?

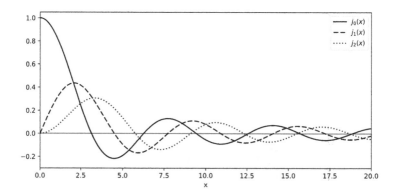

2. (a) Write a function that simulates the rolling of n dice. Use the NumPy function random.random_integers(6), which generates a random integer between 1 and 6 with equal probability (like rolling fair dice). The input of your function should be the number of dice thrown each roll and the output should be

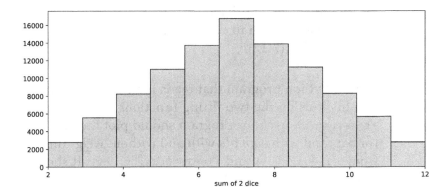

the sum of the n dice. See §9.2 for a description of NumPy's module random, which has a number of useful functions for generating arrays of random numbers.

(b) "Roll" 2 dice 10,000 times keeping track of all the sums of each set of rolls in a list. Then use your program to generate a histogram summarizing the rolls of two dice 10,000 times. The result should look like the histogram plotted above. Use the matplotlib function hist (see http://matplotlib.org/api/pyplot_summary.html) and set the number of bins in the histogram equal to the number of different possible outcomes of a roll of your dice. For example, the sum of two dice can be anything between 2 and 12, which corresponds to 11 possible outcomes. You should get a histogram that looks like the one above.

(c) Repeat part (b) using 3 dice and plot the resulting histogram.

3. In §7.5, we showed that the best fit of a line $y = a + bx$ to a set of data $\{(x_i, y_i)\}$ is obtained for the values of a and b given by Eq. (7.10). Those formulas were obtained by finding the values of a and b that minimized the sum in Eq. (7.5). This approach and these formulas are valid when the uncertainties in the data are the same for all data points. The Python function lineFit(x, y) in §7.5.1 implements Eq. (7.10).

(a) Write a new fitting function lineFitWt(x, y) that implements the formulas given in Eq. (7.14)) that minimize the χ^2 function give by Eq. (7.12). This more general approach is valid when the individual data points have different weightings *or* when they all have the same weighting. You should also write a function to calculate the reduced chi-squared χ_r^2 defined by Eq. (7.12).

(b) Write a Python program that reads in the data below, plots it, and fits it using the two fitting functions `lineFit(x, y)` and `lineFitWt(x, y)`. Your program should plot the data with error bars and with *both* fits with and without weighting, that is from `lineFit(x, y)` and `lineFitWt(x, y, dy)`. It should also report the results for both fits on the plot, similar to Fig. 7.3, as well as the values of χ_r^2, the reduce chi-squared value, for both fits. Explain why weighting the data gives a steeper or less steep slope than the fit without weighting.

```
Velocity  vs time datafor a falling  mass
time (s)  velocity (m/s) uncertainty (m/s)
  2.23        139              16
  4.78        123              16
  7.21        115               4
  9.37         96               9
 11.64         62              17
 14.23         54              17
 16.55         10              12
 18.70         -3              15
 21.05        -13              18
 23.21        -55              10
```

4. Modify the function `lineFitWt(x, y)` that you wrote in Exercise 4 above so that in addition to returning the fitting parameters a and b, it also returns the uncertainties in the fitting parameters σ_a and σ_b using the formulas given by Eq. (7.16). Use your new fitting function to find the uncertainties in the fitted slope and y-intercept for the data provided with Exercise 4.

5. Write a function to that numerically estimates the integral

$$A = \int_a^b f(x)\,dx$$

using the *trapezoid rule*. The simplest version of the trapezoid rule, which generally gives a very crude estimate, is

$$A_0 = \tfrac{1}{2}h_0[f(a) + f(b)], \quad h_0 = b - a.$$

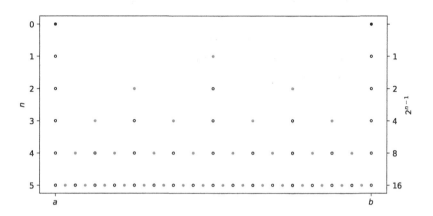

This estimate for the integral can be refined by dividing the interval from a to b in two and performing the trapezoid rule on each interval. This process can be repeated as many times as needed until you get the desired precision, which you can estimate by requiring that the fractional difference between successive estimates $(A_i - A_{i-1})/A_i < \epsilon$, where ϵ might be some small number like 10^{-8}. Repeatedly applying the trapezoid rule gives the following succession of estimates

$$A_1 = \tfrac{1}{2}h_1[f(a)+f(a+h_1)] + \tfrac{1}{2}h_1[f(a+h_1)+f(b)], \quad h_1 = \tfrac{1}{2}h_0$$
$$= \tfrac{1}{2}h_1[f(a)+2f(a+h_1)+f(b)]$$
$$= \tfrac{1}{2}A_0 + h_1 f(a+h_1)$$
$$A_2 = \tfrac{1}{2}A_1 + h_2[f(a+h_2)+f(b-h_2)], \quad h_2 = \tfrac{1}{2}h_1$$
$$A_3 = \tfrac{1}{2}A_2 + h_3[f(a+h_3)+f(a+3h_3)+f(a+5h_5)+f(b-h_3)],$$
$$h_3 = \tfrac{1}{2}h_2$$
$$\vdots$$
$$A_n = \tfrac{1}{2}A_{n-1} + h_n \sum_{i=1,3,\ldots}^{2^{n-1}} f(a+ih_n), \quad h_n = \tfrac{1}{2}h_{n-1}, \text{ for } n \geq 1$$

Write a function that implements the trapezoid rule by first evaluating A_0, then A_1, ... until ϵ is less than some preset tolerance. Note that to calculate A_i, by using the previous result A_{i-1}, you need only to evaluate the function to be integrated $f(x)$ at the open circles in the preceding diagram, saving a great deal of computation.

Try your trapezoid integration function on the following integrals and show that you get an answer within the specified tolerance of the exact value.

(a) $\int_2^5 x^2 \, dx = 39$

(b) $\int_0^\pi \sin x \, dx = 2$

(c) $\int_0^{3.5} e^{-x^2} \, dx = \frac{\sqrt{\pi}}{2} \text{erf}(3.5) \simeq 0.8862262668989721$

CHAPTER 8

Curve Fitting

> *In this chapter you learn how to use Python to perform **linear** and **nonlinear least squares fitting** of data to a function that is supposed to model the data. The methods employed allow for the weighting of the data according to uncertainties supplied by the user and uses χ^2 as the measure of the **goodness of the fit**.*

One of the most important tasks in any experimental science is modeling data and determining how well some theoretical function describes experimental data. In the last chapter, we illustrated how this can be done when the theoretical function is a simple straight line in the context of learning about Python functions and methods. Here we show how this can be done for an arbitrary fitting functions, including linear, exponential, power law, and other nonlinear fitting functions.

8.1 Using Linear Regression for Fitting Nonlinear Functions

We can use our results for linear regression with χ^2 weighting that we developed in Chapter 7 to fit functions that are nonlinear in the fitting parameters, *provided* we can transform the fitting function into one that is linear in the fitting parameters and in the independent variable (x).

8.1.1 Linear regression for fitting an exponential function

To illustrate this approach, let's consider some experimental data taken from a radioactive source that was emitting beta particles (electrons). We notice that the number of electrons emitted per unit time is decreasing with time. Theory suggests that the number of electrons N emitted per unit time should decay exponentially according to the equation

$$N(t) = N_0 e^{-t/\tau}. \tag{8.1}$$

This equation is nonlinear in t and in the fitting parameter τ and thus cannot be fit using the method of the previous chapter. Fortunately, this is a special case for which the fitting function can be transformed into a linear form. Doing so will allow us to use the fitting routine we developed for fitting linear functions.

We begin our analysis by transforming our fitting function to a linear form. To this end we take the logarithm of Eq. (8.1):

$$\ln N = \ln N_0 - \frac{t}{\tau}. \tag{8.2}$$

With this transformation, our fitting function is linear in the independent variable t. To make our method work, however, our fitting function must be linear in the *fitting parameters*, and our transformed function is still nonlinear in the fitting parameters τ and N_0. Therefore, we define new fitting parameters as follows:

$$a = \ln N_0 \tag{8.3a}$$
$$b = -1/\tau \tag{8.3b}$$

Now if we define a new dependent variable $y = \ln N$, then our fitting function takes the form of a fitting function that is linear in the fitting parameters a and b

$$y = a + bx \tag{8.4}$$

where the independent variable is $x = t$ and the dependent variable is $y = \ln N$.

We are almost ready to fit our transformed fitting function, with transformed fitting parameters a and b, to our transformed independent and dependent data, x and y. The last thing we have to do is to transform the estimates of the uncertainties δN in N to the uncertainties δy in $y (= \ln N)$. So how much does a given uncertainty in N translate into an uncertainty in y? In most cases, the uncertainty in y is much smaller than y, i.e., $\delta y \ll y$; similarly $\delta N \ll N$. In this limit we can use differentials to figure out the relationship between these

Curve Fitting

uncertainties. Here is how it works for this example:

$$y = \ln N \tag{8.5}$$

$$\delta y = \left|\frac{\partial y}{\partial N}\right| \delta N \tag{8.6}$$

$$\delta y = \frac{\partial N}{N}. \tag{8.7}$$

Equation (8.6) tells us how a small change δN in N produces a small change δy in y. Here we identify the differentials dy and dN with the uncertainties δy and δN. Therefore, an uncertainty of δN in N corresponds, or translates, to an uncertainty δy in y.

Let's summarize what we have done so far. We started with some data points $\{t_i, N_i\}$ and some addition data $\{\delta N_i\}$ where each datum δN_i corresponds to the uncertainty in the experimentally measured N_i. We wish to fit these data to the fitting function

$$N(t) = N_0 e^{-t/\tau}.$$

We then take the natural logarithm of both sides and obtain the linear equation

$$\ln N = \ln N_0 - \frac{t}{\tau} \tag{8.8a}$$

$$y = a + bx \tag{8.8b}$$

with the obvious correspondences

$$x = t \tag{8.9a}$$
$$y = \ln N \tag{8.9b}$$
$$a = \ln N_0 \tag{8.9c}$$
$$b = -1/\tau. \tag{8.9d}$$

Now we can use the linear regression routine with χ^2 weighting that we developed in the previous section to fit Eq. (8.4) to the transformed data $x_i (= t_i)$ and $y_i (= \ln N_i)$. The inputs are the transformed data: x_i, y_i, and δy_i. The outputs are the fitting parameters a and b, as well as the estimates of their uncertainties δa and δb along with the value of χ^2. You can obtain the uncertainties δN_0 and $\delta \tau$ of the original fitting parameters N_0 and τ by taking the differentials of Eqs. (8.9c) and (8.9d):

$$\delta a = \left|\frac{\partial a}{\partial N_0}\right| \delta N_0 = \frac{\delta N_0}{N_0} \tag{8.10}$$

$$\delta b = \left|\frac{\partial b}{\partial \tau}\right| \delta \tau = \frac{\delta \tau}{\tau^2} \tag{8.11}$$

The Python routine below shows how to implement all of this for a set of experimental data that is read in from a data file.

Figure 8.1 shows the output of the fit to simulated beta decay data obtained using the program below. Note that the error bars are large when the number of counts N are small. This is consistent with what is known as *shot noise* (noise that arises from counting discrete events), which obeys *Poisson* statistics. The program also prints out the fitting parameters of the transformed data as well as the fitting parameters for the exponential fitting function.

Code: chapter8/programs/betaDecay.py

```
import numpy as np
import matplotlib.pyplot as plt


def LineFitWt(x, y, dy):
    """
    Fit to straight line.
    Inputs: x, y, and dy (y-uncertainty) arrays.
    Ouputs: slope and y-intercept of best fit to data.
    """
    dy2 = dy**2
    norm = (1./dy2).sum()
    xhat = (x/dy2).sum() / norm
    yhat = (y/dy2).sum() / norm
    slope = ((x-xhat)*y/dy2).sum()/((x-xhat)*x/dy2).sum()
    yint = yhat - slope*xhat
    dy2_slope = 1./((x-xhat)*x/dy2).sum()
    dy2_yint = dy2_slope * (x*x/dy2).sum() / norm
    return slope, yint, np.sqrt(dy2_slope), np.sqrt(dy2_yint)


def redchisq(x, y, dy, slope, yint):
    chisq = (((y-yint-slope*x)/dy)**2).sum()
    return chisq/float(x.size-2)



# Read data from data file
t, N, dN = np.loadtxt("betaDecay.txt", skiprows=2, unpack=True)
```

Curve Fitting

```python
# Transform data and parameters to linear form: Y = A + B*X
X = t              # transform t data for fitting (trivial)
Y = np.log(N)      # transform N data for fitting
dY = dN/N          # transform uncertainties for fitting

# Fit transformed data X, Y, dY --> fitting parameters A & B
# Also returns uncertainties in A and B
B, A, dB, dA = LineFitWt(X, Y, dY)
# Return reduced chi-squared
redchisqr = redchisq(X, Y, dY, B, A)

# Determine fitting parameters for exponential function
# N = N0 exp(-t/tau) ...
N0 = np.exp(A)
tau = -1.0/B
# ... and their uncertainties
dN0 = N0 * dA
dtau = tau**2 * dB

# Code to plot transformed data and fit starts here
# Create line corresponding to fit using fitting parameters
# Only two points are needed to specify a straight line
Xext = 0.05*(X.max()-X.min())
Xfit = np.array([X.min()-Xext, X.max()+Xext])
Yfit = A + B*Xfit

fig, ax = plt.subplots()
ax.errorbar(X, Y, dY, fmt="oC0")
ax.plot(Xfit, Yfit, "-C1", zorder=-1)
ax.set_xlim(0, 100)
ax.set_ylim(1.5, 7)
ax.set_title(r"$\mathrm{Fit\ to:}\ \ln N = -t/\tau + \ln N_0$")
ax.set_xlabel("$t$")
ax.set_ylabel("ln($N$)")
ax.text(50, 6.6, "A = ln N0 = {0:0.2f} $\pm$ {1:0.2f}"
        .format(A, dA))
ax.text(50, 6.3, "B = -1/tau = {0:0.4f} $\pm$ {1:0.4f}"
        .format(-B, dB))
ax.text(50, 6.0, "$\chi_r^2$ = {0:0.3f}"
        .format(redchisqr))
ax.text(50, 5.7, "N0 = {0:0.0f} $\pm$ {1:0.0f}"
        .format(N0, dN0))
ax.text(50, 5.4, "tau = {0:0.1f} $\pm$ {1:0.1f} days"
        .format(tau, dtau))
fig.savefig("figures/betaDecay.pdf")
fig.show()
```

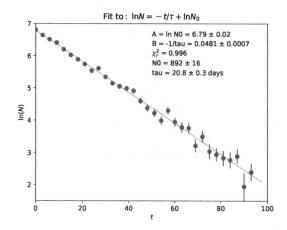

Figure 8.1 Semi-log plot of beta decay measurements from Phosphorus-32.

8.1.2 Linear regression for fitting a power-law function

You can use a similar approach to the one outlined above to fit experimental data to a power law fitting function of the form

$$P(s) = P_0 s^\alpha. \tag{8.12}$$

We follow the same approach we used for the exponential fitting function and first take the logarithm of both sides of Eq. (8.12)

$$\ln P = \ln P_0 + \alpha \ln s. \tag{8.13}$$

We recast this in the form of a linear equation $y = a + bx$ with the following identifications:

$$x = \ln s \tag{8.14a}$$
$$y = \ln P \tag{8.14b}$$
$$a = \ln P_0 \tag{8.14c}$$
$$b = \alpha \tag{8.14d}$$

Following a procedure similar to that used to fit using an exponential fitting function, you can use the transformations given by Eq. (8.14) as the basis for a program to fit a power-law fitting function such as Eq. (8.13) to experimental data.

Curve Fitting

8.2 Nonlinear Fitting

The method introduced in the previous section for fitting nonlinear fitting functions can be used only if the fitting function can be transformed into a fitting function that is linear in the fitting parameters $a, b, c...$ When we have a nonlinear fitting function that cannot be transformed into a linear form, we need another approach.

The problem of finding values of the fitting parameters that minimize χ^2 is a nonlinear optimization problem to which there is quite generally no analytical solution (in contrast to the linear optimization problem). We can gain some insight into this nonlinear optimization problem, namely the fitting of a nonlinear fitting function to a data set, by considering a fitting function with only two fitting parameters. That is, we are trying to fit some data set $\{x_i, y_i\}$, with uncertainties in $\{y_i\}$ of $\{\sigma_i\}$, to a fitting function $f(x; a, b)$ where a and b are the two fitting parameters. To do so, we look for the minimum in

$$\chi^2(a,b) = \sum_i \left(\frac{y_i - f(x_i)}{\sigma_i} \right). \tag{8.15}$$

Note that once the data set, uncertainties, and fitting function are specified, $\chi^2(a, b)$ is simply a function of a and b. We can picture the function $\chi^2(a, b)$ as a landscape with peaks and valleys: as we vary a and b, $\chi^2(a, b)$ rises and falls. The basic idea of all nonlinear fitting routines is to start with some initial guesses for the fitting parameters, here a and b, and by scanning the $\chi^2(a, b)$ landscape, find values of a and b that minimize $\chi^2(a, b)$.

There are a number of different methods for trying to find the minimum in χ^2 for nonlinear fitting problems. Nevertheless, the method that is most widely used goes by the name of the *Levenberg-Marquardt* method. Actually, the Levenberg-Marquardt method is a combination of two other methods, the *steepest descent* (or gradient) method and *parabolic extrapolation*. Roughly speaking, when the values of a and b are not too near their optimal values, the gradient descent method determines in which direction in (a, b)-space the function $\chi^2(a, b)$ decreases most quickly—the direction of steepest descent—and then changes a and b to move in that direction. This method is very efficient unless a and b are very near their optimal values. Near the optimal values of a and b, parabolic extrapolation is more efficient. Therefore, as a and b approach their optimal values, the Levenberg-Marquardt method gradually changes to the parabolic extrapolation method, which approximates $\chi^2(a,b)$ by a Taylor series second-order in a and b and then computes directly the analytical minimum of the Taylor series approximation of $\chi^2(a,b)$. This method is only good if the second-order Taylor series provides a good approximation of $\chi^2(a,b)$. That is why parabolic extrapolation only works well very near the minimum in $\chi^2(a,b)$.

Before illustrating the Levenberg-Marquardt method, we make one important cautionary remark: the Levenberg-Marquardt method can fail if the initial guesses of the fitting parameters are too far away from the desired solution. This problem becomes more serious the greater the number of fitting parameters. Thus it is important to provide reasonable initial guesses for the fitting parameters. Usually, this is not a problem, as it is clear from the physical situation of a particular experiment what reasonable values of the fitting parameters are. But beware!

The `scipy.optimize` module provides routines that implement the Levenberg-Marquardt nonlinear fitting method. One most useful of these is called `scipy.optimize.curve_fit` and it is the one we demonstrate here. The function call is

```
import scipy.optimize
[... insert code here ...]
scipy.optimize.curve_fit(f, xdata, ydata, p0=None,
                         sigma=None, **kwargs)
```

The arguments of `curve_fit` are as follows:

f(xdata, a, b, ...): is the fitting function where `xdata` is the data for the independent variable and `a`, `b`, ... are the fitting parameters, however many there are, listed as separate arguments. Obviously, `f(xdata, a, b, ...)` should return the y value of the fitting function.

xdata: is the array containing the x data.

ydata: is the array containing the y data.

p0: is a tuple containing the initial guesses for the fitting parameters. The guesses for the fitting parameters are set equal to 1 if they are left unspecified. It is almost always a good idea to specify the initial guesses for the fitting parameters.

sigma: is the array containing the uncertainties in the y data.

Curve Fitting

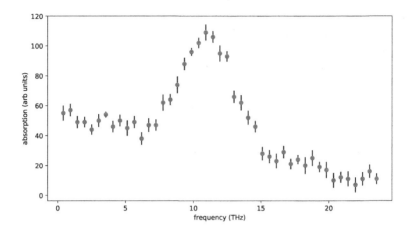

Figure 8.2 Data.

kwargs: are keyword arguments that can be passed to the fitting routine `scipy.optimize.leastsq` that `curve_fit` calls. These are usually left unspecified.

We demonstrate the use of `curve_fit` to fit the data plotted in Fig. 8.2. We model the data with the fitting function that consists of a quadratic polynomial background with a Gaussian peak:

$$s(f) = a + bf + cf^2 + Pe^{-\frac{1}{2}[(f-f_p)/f_w]^2} \tag{8.16}$$

Lines 8 and 9 define the fitting function. Note that the independent variable `f` is the first argument, which is followed by the six fitting parameters a, b, c, P, f_p, and f_w.

To fit the data with $s(f)$, we need good estimates of the fitting parameters. Setting $f = 0$, we see that $a \approx 60$. An estimate of the slope of the baseline gives $b \approx -60/20 = -3$. The curvature in the baseline is small so we take $c \approx 0$. The amplitude of the peak above the baseline is $P \approx 80$. The peak is centered at $f_p \approx 11$, while the width of the peak is about $f_w \approx 2$. We use these estimates to set the initial guesses of the fitting parameters in lines 18 and 19 in the following code.

The function that performs the Levenberg-Marquardt algorithm, `scipy.optimize.curve_fit`, is called in lines 22–23 with the output set equal to the one- and two-dimensional arrays `nlfit` and `nlpcov`, respectively. The array `nlfit`, which gives the optimal values of the fitting parameters, is unpacked in line 26. The square root of the diagonal of the two-dimensional array `nlpcov`, which gives the estimates of the uncertainties in the fitting parameters, is unpacked in lines 29–30 using a list comprehension.

The rest of the code plots the data, the fitting function using the optimal values of the fitting parameters found by `scipy.optimize.curve_fit`, and the values of the fitting parameters and their uncertainties.

Code: chapter8/programs/fitSpectrum1.py

```python
import numpy as np
import matplotlib.pyplot as plt
import matplotlib.gridspec as gridspec   # unequal plots
import scipy.optimize


# define fitting function
def GaussPolyBase(f, a, b, c, P, fp, fw):
    return a+b*f+c*f*f+P*np.exp(-0.5*((f-fp)/fw)**2)


# read in spectrum from data file
# f=frequency, s=signal, ds=s uncertainty
f, s, ds = np.loadtxt("Spectrum.txt", skiprows=4,
                      unpack=True)

# initial guesses for fitting parameters
a0, b0, c0 = 60., -3., 0.
P0, fp0, fw0 = 80., 11., 2.

# fit data using SciPy's Levenberg Marquart method
nlfit, nlpcov = scipy.optimize.curve_fit(GaussPolyBase,
        f, s, p0=[a0, b0, c0, P0, fp0, fw0], sigma=ds)

# unpack fitting parameters
a, b, c, P, fp, fw = nlfit
# unpack uncertainties in fitting parameters from
# diagonal of covariance matrix
da, db, dc, dP, dfp, dfw = [np.sqrt(nlpcov[j, j])
                            for j in range(nlfit.size)]

# create fitting function from fitted parameters
f_fit = np.linspace(0.0, 25., 128)
s_fit = GaussPolyBase(f_fit, a, b, c, P, fp, fw)

# Calculate residuals and reduced chi squared
resids = s - GaussPolyBase(f, a, b, c, P, fp, fw)
```

Curve Fitting

```
38    redchisqr = ((resids/ds)**2).sum()/float(f.size-6)
39
40    # Create figure window to plot data
41    fig = plt.figure(1, figsize=(9.5, 6.5))
42    gs = gridspec.GridSpec(2, 1, height_ratios=[6, 2])
43
44    # Top plot: data and fit
45    ax1 = fig.add_subplot(gs[0])
46    ax1.plot(f_fit, s_fit, '-C0')
47    ax1.errorbar(f, s, yerr=ds, fmt='oC3', ecolor='black')
48    ax1.set_xlabel('frequency (THz)')
49    ax1.set_ylabel('absorption (arb units)')
50    ax1.text(0.7, 0.95, 'a = {0:0.1f}$\pm${1:0.1f}'
51            .format(a, da), transform=ax1.transAxes)
52    ax1.text(0.7, 0.90, 'b = {0:0.2f}$\pm${1:0.2f}'
53            .format(b, db), transform=ax1.transAxes)
54    ax1.text(0.7, 0.85, 'c = {0:0.2f}$\pm${1:0.2f}'
55            .format(c, dc), transform=ax1.transAxes)
56    ax1.text(0.7, 0.80, 'P = {0:0.1f}$\pm${1:0.1f}'
57            .format(P, dP), transform=ax1.transAxes)
58    ax1.text(0.7, 0.75, 'fp = {0:0.1f}$\pm${1:0.1f}'
59            .format(fp, dfp), transform=ax1.transAxes)
60    ax1.text(0.7, 0.70, 'fw = {0:0.1f}$\pm${1:0.1f}'
61            .format(fw, dfw), transform=ax1.transAxes)
62    ax1.text(0.7, 0.60, '$\chi_r^2$ = {0:0.2f}'
63            .format(redchisqr), transform=ax1.transAxes)
64    ax1.set_title('$s(f)=a+bf+cf^2+P\,e^{-(f-f_p)^2/2f_w^2}$')
65
66    # Bottom plot: residuals
67    ax2 = fig.add_subplot(gs[1])
68    ax2.errorbar(f, resids, yerr=ds, ecolor="black",
69                 fmt="oC3")
70    ax2.axhline(color="gray", zorder=-1)
71    ax2.set_xlabel('frequency (THz)')
72    ax2.set_ylabel('residuals')
73    ax2.set_ylim(-20, 20)
74    ax2.set_yticks((-20, 0, 20))
75
76    fig.savefig("figures/fitSpectrum.pdf")
77    fig.show()
```

The above code also plots the difference between the data and fit, known as the *residuals*, in the subplot below the plot of the data and fit. Plotting the residuals in this way gives a graphical representation of the goodness of the fit. To the extent that the residuals vary randomly about zero and do not show any overall upward or downward curvature, or any long wavelength oscillations, the fit would seem to be a good fit.

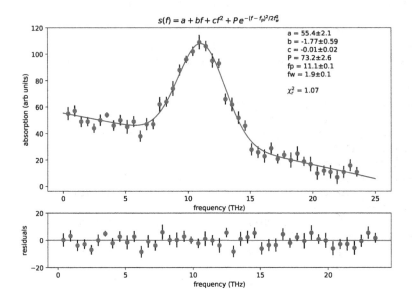

Figure 8.3 Fit to Gaussian with quadratic polynomial background.

Finally, we note that we have used the matplotlib package `gridspec` to create the two subplots with different heights. The `gridspec` are made in lines 3 (where the package is imported), 42 (where 2 rows and 1 column are specified with relative heights of 6 to 2), 45 (where the first `gs[0]` height is specified), and 67 (where the second lstinlinegs[1] height is specified). More details about the `gridspec` package can be found at the matplotlib web site.

8.3 Exercises

1. Fit the data below to a straight line using the function `LineFitWt(x, y, dy)` introduced in the program `betaDecay.py` on page 190. Plot the data and the fit so that you obtain a figure like the one below, which includes the best fit (solid line) along with two fits shown with dashed lines. One has a slope equal to the fitted slope *plus* the slope uncertainty obtained from the fit. The other has a slope equal to the fitted slope *minus* the slope uncer-

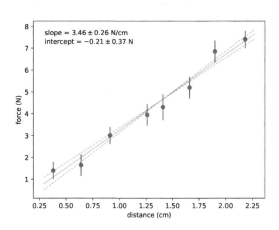

tainty obtained from the fit. These are matched with the fitted y-intercept minus or plus its fitted uncertainty.

```
d = np.array([0.38, 0.64, 0.91, 1.26, 1.41, 1.66, 1.90,
              2.18])
f = np.array([1.4, 1.65, 3.0, 3.95, 4.3, 5.20, 6.85, 7.4])
df = np.array([0.4, 0.5, 0.4, 0.5, 0.6, 0.5, 0.5, 0.4])
```

2. When a voltage source is connected across a resistor and inductor in series, the voltage across the inductor $V_i(t)$ is predicted to obey the equation

$$V(t) = V_0 e^{-\Gamma t} \qquad (8.17)$$

where t is the time and the decay rate $\Gamma = R/L$ is the ratio of the resistance R to the inductance L of the circuit. In this problem, you are to write a Python routine that fits the above equation to the data below for the voltage measured across an inductor after it is connected in series with a resistor to a voltage source. Following the example in the text, linearize Eq. (8.17) and use a linear fitting routine, either the one you wrote from the previous chapter or one from NumPy or SciPy.

(a) Find the best values of Γ and V_0 and the uncertainties in their values σ_Γ and σV_0.

(b) Find the value of χ_r^2 for your fit. Does it make sense?

(c) Make a semi-log plot of the data using symbols with error bars (no line) and of the fit (line only). The fit should appear as a straight line that goes through the data points.

```
Data for decay of voltage across an inductor
in an RL circuit
Date: 24-Oct-2012
Data taken by D. M. Blantogg and T. P. Chaitor

time (ns)    voltage (volts)    uncertainty (volts)
   0.0           5.08e+00            1.12e-01
  32.8           3.29e+00            9.04e-02
  65.6           2.23e+00            7.43e-02
  98.4           1.48e+00            6.05e-02
 131.2           1.11e+00            5.25e-02
 164.0           6.44e-01            4.00e-02
 196.8           4.76e-01            3.43e-02
 229.6           2.73e-01            2.60e-02
 262.4           1.88e-01            2.16e-02
 295.2           1.41e-01            1.87e-02
 328.0           9.42e-02            1.53e-02
 360.8           7.68e-02            1.38e-02
 393.6           3.22e-02            8.94e-03
 426.4           3.22e-02            8.94e-03
 459.2           1.98e-02            7.01e-03
 492.0           1.98e-02            7.01e-03
```

3. Fit the data of the previous exercise to fit Eq. (8.17) using the SciPy function `scipy.optimize.curve_fit`. Plot the data as symbols and the fit as a line on linear and on semilogarithmic axes in two separate plots in the same figure window. Compare your results to those of the previous exercise.

4. Small nanoparticles of soot suspended in water start to aggregate when salt is added. The average radius r of the aggregates is predicted to grow as a power law in time t according to the equation

$$r = Kt^p. \tag{8.18}$$

In general the power p is not an integer, which means that K has odd units that depend on the value of p. Taking the logarithm of this equation gives $\ln r = p \ln t + \ln K$. Thus, the data should fall on a straight line of slope p if $\ln r$ is plotted vs. $\ln t$.

   (a) Plot the data below on a graph of $\ln r$ vs. $\ln t$ to see if the data fall approximately on a straight line.

Curve Fitting

```
Size of growing aggregate
Date: 19-Nov-2013
Data taken by M. D. Gryart and A. D. Waites
time (m)      size (nm)      unc (nm)
 0.12           115             10
 0.18           130             12
 0.42           202             14
 0.90           335             18
 2.10           510             20
 6.00           890             30
18.00          1700             40
42.00          2600             50
```

(b) Defining $y = \ln r$ and $x = \ln t$, use the linear fitting routine you wrote for the previous problem to fit the data and find the optimal values for the slope and y intercept, as well as their uncertainties. Use these fitted values to find the optimal values of the amplitude K and the power p in the fitting function $r = K t^p$. What are the fitted values of K and p and their uncertainties? What is the value of χ^2? Does a power law provide an adequate model for the data?

5. Fit the data of the previous exercise to fit Eq. (8.18) using the SciPy function `scipy.optimize.curve_fit`. Plot the data as symbols and the fit as a line on linear and on log-log axes in two separate plots in the same figure window. Compare your results to those of the previous exercise.

6. In this problem you explore using a nonlinear least square fitting routine to fit the data shown in the figure on the next page. The data, including the uncertainties in the y values, are provided in the table at the end of this problem. Your task is to fit the function

$$d(t) = A(1 + B\cos\omega t)e^{-t^2/2\tau^2} + C \tag{8.19}$$

to the data, where the fitting parameters are A, B, C, ω, and τ.

(a) Write a Python program that (i) reads the data in from a data file, (ii) defines a function `oscDecay(t, A, B, C, tau, omega)` for the function $d(t)$ above, and (iii) produces a plot of the data and the function $d(t)$. Choose the fitting parameters A, B, C, tau, and omega to produce an approximate fit "by eye" to the data. You should be able estimate reasonable values for these parameters just by looking at the data and thinking

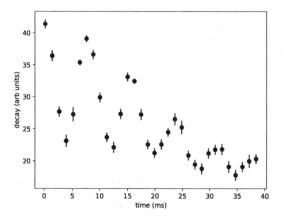

about the behavior of $d(t)$. For example, $d(0) = A(1 + B) + C$ while $d(\infty) = C$. What parameter in $d(t)$ controls the period of the peaks observed in the data? Use that information to estimate the value of that parameter.

(b) Following the example in §8.2, write a program using the SciPy function `scipy.optimize.curve_fit` to fit Eq. (8.19) to the data and thus find the optimal values of the fitting parameters A, B, C, ω, and τ. Your program should plot the data along with the fitting function using the optimal values of the fitting parameters. Write a function to calculate the reduced χ^2. Print out the value of the reduced χ^2 on your plot along with the optimal values of the fitting parameters. You can use the results from part (a) to estimate good starting values of the fitting parameters.

(c) Once you have found the optimal fitting parameters, run your fitting program again using for starting values the optimal values of the fitting parameters A, B, C, and τ, but set the starting value of ω to be 3 times the optimal value. You should find that the program converges to a different set of fitting parameters than the ones you found in part (b). Using the program you wrote for part (b), make a plot of the data and the fit like the one you did for part (a). The fit should be noticeably worse. What is the value of the reduced χ^2 for this fit? It should be much larger than the one you found for part

Curve Fitting

(c). The program has found a local minimum in χ^2—one that obviously is not the best fit!

(d) Setting the fitting parameters A, B, C, and τ to the optimal values you found in part (b), plot χ_r^2 as a function of ω for ω spanning the range from 0.05 to 3.95. You should observe several local minima for different values of χ_r^2; the global minimum in χ_r^2 should occur for the optimal value of ω you found in part (b).

```
Data for absorption spectrum
Date: 21-Nov-2012
Data taken by P. Dubson and M. Skraps
time (ms)      signal     uncertainty
  0.2           41.1          0.9
  1.4           37.2          0.9
  2.7           28.3          0.9
  3.9           24.8          1.1
  5.1           27.8          0.8
  6.4           34.5          0.7
  7.6           39.0          0.9
  8.8           37.7          0.8
 10.1           29.8          0.9
 11.3           22.2          0.7
 12.5           22.3          0.6
 13.8           26.7          1.1
 15.0           30.4          0.7
 16.2           32.6          0.8
 17.5           28.9          0.8
 18.7           22.9          1.3
 19.9           21.7          0.9
 21.1           22.1          1.0
 22.4           22.3          1.0
 23.6           26.3          1.0
 24.8           26.2          0.8
 26.1           21.4          0.9
 27.3           20.0          1.0
 28.5           20.1          1.2
 29.8           21.2          0.5
 31.0           22.0          0.9
 32.2           21.6          0.7
 33.5           21.0          0.7
 34.7           19.7          0.9
 35.9           17.9          0.9
 37.2           18.1          0.8
 38.4           18.9          1.1
```

CHAPTER 9

Numerical Routines: SciPy and NumPy

> *This chapter describes some of the more useful numerical routines available in the SciPy and NumPy packages, most of which are wrappers to well-established numerical routines written in Fortran, C, and C++.* **Random number generators** *are covered. Linear algebra routines are covered, including ones that solve* **systems of linear equations** *and* **eigenvalue problems***. Routines for obtaining* **solutions to nonlinear equations** *are introduced, as are routines to perform* **numerical integration** *of both single and multiple integrals. Routines for obtaining* **solutions to ODEs** *(and systems of ODEs) are introduced. Finally, you learn about routines to perform* **discrete Fourier transforms** *(FFT algorithm).*

SciPy is a Python library of mathematical routines. Many of the SciPy routines are Python "wrappers," that is, Python routines that provide a Python interface, for numerical libraries and routines originally written in Fortran, C, or C++. Thus, SciPy lets you take advantage of the decades of work that has gone into creating and optimizing numerical routines for science and engineering. Because the Fortran, C, or C++ code that Python accesses is compiled, these routines typically run very fast. Therefore, there is no real downside—no speed penalty—for using Python in these cases.

We already encountered SciPy's routine for fitting nonlinear functions to experimental data, `scipy.optimize.leastsq`, which was introduced in Chapter 8. Here we will provide a further introduction to a number of other SciPy packages, in particular those on special functions, linear algebra, finding roots of scalar functions, discrete Fourier transforms, and numerical integration, including routines for numerically solving ordinary differential equations (ODEs). Our introduction to these capabilities does not include extensive background on the numerical methods employed; that is a topic for another text. Here we simply introduce the SciPy routines for performing some of the more frequently required numerical tasks.

One final note: SciPy makes extensive use of NumPy arrays, so NumPy should be imported with SciPy.

9.1 Special Functions

SciPy provides a plethora of special functions, including Bessel functions (and routines for finding their zeros, derivatives, and integrals), error functions, the gamma function, Legendre, Laguerre, and Hermite polynomials (and other polynomial functions), Mathieu functions, many statistical functions, and a number of other functions. Most are contained in the scipi.special library, and each has its own special arguments and syntax, depending on the vagaries of the particular function. We demonstrate a number of them in the code below that produces a plot of the different functions called. For more information, you should consult the SciPy web site on the scipy.special library.

Code: chapter9/programs/specFuncPlotsBW.py

```
import numpy as np
import scipy.special
import matplotlib.pyplot as plt

# create a figure window with subplots
fig, ax = plt.subplots(3, 2, figsize=(9.3, 6.5))


# create arrays for a few Bessel functions and plot them
x = np.linspace(0, 20, 256)
j0 = scipy.special.jn(0, x)
j1 = scipy.special.jn(1, x)
y0 = scipy.special.yn(0, x)
y1 = scipy.special.yn(1, x)
ax[0, 0].plot(x, j0, color='black')
ax[0, 0].plot(x, j1, color='black', dashes=(5, 2))
ax[0, 0].plot(x, y0, color='black', dashes=(3, 2))
ax[0, 0].plot(x, y1, color='black', dashes=(1, 2))
ax[0, 0].axhline(color="grey", ls="--", zorder=-1)
ax[0, 0].set_ylim(-1, 1)
ax[0, 0].text(0.5, 0.95, 'Bessel', ha='center',
              va='top', transform=ax[0, 0].transAxes)

# gamma function
x = np.linspace(-3.5, 6., 3601)
g = scipy.special.gamma(x)
g = np.ma.masked_outside(g, -100, 400)
```

Numerical Routines: SciPy and NumPy

```
27  ax[0, 1].plot(x, g, color='black')
28  ax[0, 1].set_xlim(-3.5, 6)
29  ax[0, 1].axhline(color="grey", ls="--", zorder=-1)
30  ax[0, 1].axvline(color="grey", ls="--", zorder=-1)
31  ax[0, 1].set_ylim(-20, 100)
32  ax[0, 1].text(0.5, 0.95, 'Gamma', ha='center',
33                va='top', transform=ax[0, 1].transAxes)
34
35  # error function
36  x = np.linspace(0, 2.5, 256)
37  ef = scipy.special.erf(x)
38  ax[1, 0].plot(x, ef, color='black')
39  ax[1, 0].set_ylim(0, 1.1)
40  ax[1, 0].text(0.5, 0.95, 'Error', ha='center',
41                va='top', transform=ax[1, 0].transAxes)
42
43  # Airy function
44  x = np.linspace(-15, 4, 256)
45  ai, aip, bi, bip = scipy.special.airy(x)
46  ax[1, 1].plot(x, ai, color='black')
47  ax[1, 1].plot(x, bi, color='black', dashes=(5, 2))
48  ax[1, 1].axhline(color="grey", ls="--", zorder=-1)
49  ax[1, 1].axvline(color="grey", ls="--", zorder=-1)
50  ax[1, 1].set_xlim(-15, 4)
51  ax[1, 1].set_ylim(-0.5, 0.6)
52  ax[1, 1].text(0.5, 0.95, 'Airy', ha='center',
53                va='top', transform=ax[1, 1].transAxes)
54
55  # Legendre polynomials
56  x = np.linspace(-1, 1, 256)
57  lp0 = np.polyval(scipy.special.legendre(0), x)
58  lp1 = np.polyval(scipy.special.legendre(1), x)
59  lp2 = scipy.special.eval_legendre(2, x)
60  lp3 = scipy.special.eval_legendre(3, x)
61  ax[2, 0].plot(x, lp0, color='black')
62  ax[2, 0].plot(x, lp1, color='black', dashes=(5, 2))
63  ax[2, 0].plot(x, lp2, color='black', dashes=(3, 2))
64  ax[2, 0].plot(x, lp3, color='black', dashes=(1, 2))
65  ax[2, 0].axhline(color="grey", ls="--", zorder=-1)
66  ax[2, 0].axvline(color="grey", ls="--", zorder=-1)
67  ax[2, 0].set_ylim(-1, 1.1)
68  ax[2, 0].text(0.5, 0.9, 'Legendre', ha='center',
69                va='top', transform=ax[2, 0].transAxes)
70
71  # Laguerre polynomials
72  x = np.linspace(-5, 8, 256)
73  lg0 = np.polyval(scipy.special.laguerre(0), x)
74  lg1 = np.polyval(scipy.special.laguerre(1), x)
75  lg2 = scipy.special.eval_laguerre(2, x)
```

```
76  lg3 = scipy.special.eval_laguerre(3, x)
77  ax[2, 1].plot(x, lg0, color='black')
78  ax[2, 1].plot(x, lg1, color='black', dashes=(5, 2))
79  ax[2, 1].plot(x, lg2, color='black', dashes=(3, 2))
80  ax[2, 1].plot(x, lg3, color='black', dashes=(1, 2))
81  ax[2, 1].axhline(color="grey", ls="--", zorder=-1)
82  ax[2, 1].axvline(color="grey", ls="--", zorder=-1)
83  ax[2, 1].set_xlim(-5, 8)
84  ax[2, 1].set_ylim(-5, 10)
85  ax[2, 1].text(0.5, 0.9, 'Laguerre', ha='center',
86                va='top', transform=ax[2, 1].transAxes)
87  fig.tight_layout()
88  fig.savefig("specFuncPlotsBW.pdf")
89  fig.show()
```

The arguments of the different functions depend, of course, on the nature of the particular function. For example, the first argument of the two types of Bessel functions called in lines 10–13 is the so-called *order* of the Bessel function, and the second argument is the independent variable. The Gamma and Error functions take one argument each and produce one output. The Airy function takes only one input argument, but returns four outputs, which correspond the two Airy

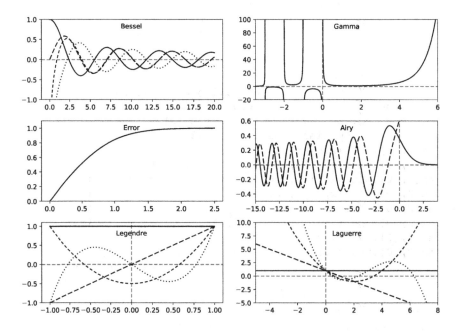

Figure 9.1 Plots of special functions.

functions, normally designated Ai(x) and Bi(x), and their derivatives Ai'(x) and Bi'(x). The plot shows only Ai(x) and Bi(x).

The polynomial functions shown have a special syntax that uses NumPy's `polyval` function for generating polynomials. If `p` is a list or array of N numbers and x is an array, then

```
polyval(p, x) = p[0]*x**(N-1) + p[1]*x**(N-2)
              + ... + p[N-2]*x + p[N-1]
```

For example, if `p = [2.0, 5.0, 1.0]`, `polyval` generates the following quadratic polynomial:

```
polyval(p, x) = 2.0*x**2 + 5.0*x + 1.0
```

SciPy's `special.legendre(n)` and `special.laguerre(n)` functions output the coefficients p needed in `polyval` to produce the nth-order Legendre and Laguerre polynomials, respectively. The `special` library of SciPy has functions that specify many other polynomial functions in this same way.

Alternatively, the Legendre polynomials can be calculated using simpler function calls `scipy.special.eval_legendre(n, x)`, where n is the order of the polynomial and x is the array of points where the function is evaluated. Similar function calls exist for Leguerre and other common polynomials. These `scipy.special.eval_*` function calls are preferred when the order n is greater than about 20, where using `polyval` starts to become unstable.

9.2 Random Numbers

Random numbers are widely used in science and engineering computations. They can be used to simulate noisy data, or to model physical phenomena like the distribution of velocities of molecules in a gas, or to act like the roll of dice in a game. There are even methods for numerically evaluating multi-dimensional integrals using random numbers.

The basic idea of a random number generator is that it should be able to produce a sequence of numbers that are distributed according to some predetermined distribution function. NumPy provides a number of such random number generators in its library `numpy.random`. Here we focus on three: `rand`, `randn`, and `randint`.

9.2.1 Uniformly distributed random numbers

The `rand(num)` function creates an array of num floats uniformly distributed on the interval from 0 to 1.

```
In [1]: rand()
Out[1]: 0.5885170150833566

In [2]: rand(5)
Out[2]: array([ 0.85586399, 0.21183612, 0.80235691,
               0.65943861, 0.25519987])
```

If `rand` has no argument, a single random number is generated. Otherwise, the argument specifies the number of random numbers (and size of the array) that is created.

If you want random numbers uniformly distributed over some other interval, say from a to b, then you can do that simply by stretching the interval so that it has a width of $b-a$ and displacing the lower limit from 0 to a. The following statements produce random numbers uniformly distributed from 10 to 20:

```
In [3]: a, b = 10, 20

In [4]: (b-a)*rand(20) + a
Out[4]: array([ 10.99031149, 18.11685555, 11.48302458,
               18.25559651, 17.55568817, 11.86290145,
               17.84258224, 12.1309852 , 14.30479884,
               12.05787676, 19.63135536, 16.58552886,
               19.15872073, 17.59104303, 11.48499468,
               10.16094915, 13.95534353, 18.21502143,
               19.61360422, 19.21058726])
```

9.2.2 Normally distributed random numbers

The function `randn(n)` produces a *normal* or *Gaussian* distribution of n random numbers with a mean of 0 and a standard deviation of 1. That is, they are distributed according to

$$P(x) = \frac{1}{\sqrt{2\pi}} e^{-\frac{1}{2}x^2}.$$

Figure 9.2 shows histograms for the distributions of 10,000 random numbers generated by the `np.random.rand` and `np.random.randn` functions. As advertised, the `np.random.rand` function produces an array of random numbers that is uniformly distributed between the values of 0 and 1, while the `np.random.randn` function produces an array of

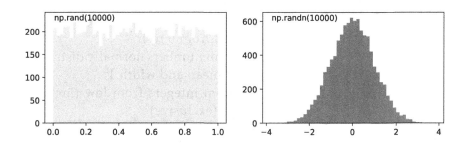

Figure 9.2 Random number distributions.

random numbers that follows a distribution of mean 0 and standard deviation 1.

If we want random numbers with a Gaussian distribution of width σ centered about x_0, we stretch the interval by a factor of σ and displace it by x_0. The following code produces 20 random numbers normally distributed around 15 with a width of 10:

```
In [5]: x0, sigma = 15, 10

In [6]: sigma*randn(20) + x0
Out[6]: array([ 9.36069244, 13.49260733,  6.12550102,
               18.50471781,  9.89499319, 14.09576728,
               12.45076637, 17.83073628,  2.95085564,
               18.2756275 , 14.781659  , 31.80264078,
               20.8457924 , 13.87890601, 25.41433678,
               15.44237582, 21.2385386 , -3.91668973,
               31.19120157, 26.24254326])
```

9.2.3 Random distribution of integers

The function `randint(low, high, n)` produces a uniform random distribution of `n` integers between `low` (inclusive) and `high` (exclusive). For example, we can simulate a dozen rolls of a single die with the following statement:

```
In [7]: randint(1, 7, 12)
Out[7]: array([6, 2, 1, 5, 4, 6, 3, 6, 5, 4, 6, 2])
```

Function call	Output
rand(n)	n random numbers uniformly distributed from 0 to 1
randn(n)	n random numbers normally distributed with 0 mean and width 1
randint(low, high, n)	n random integers from low (inclusive) to high (exclusive)

Table 9.1 Some random number functions from numpy.random.

9.3 Linear Algebra

Python's mathematical libraries, NumPy and SciPy, have extensive tools for numerically solving problems in linear algebra. Here we focus on two problems that arise commonly in scientific and engineering settings: (1) solving a system of linear equations and (2) eigenvalue problems. In addition, we show how to perform a number of other basic computations, such as finding the determinant of a matrix, matrix inversion, and *LU* decomposition. The SciPy package for linear algebra is called `scipy.linalg`.

9.3.1 Basic computations in linear algebra

SciPy has a number of routines for performing basic operations with matrices. The determinant of a matrix is computed using the `scipy.linalg.det` function:

```
In [1]: import scipy.linalg
In [2]: a = array([[-2, 3], [4, 5]])
In [3]: a
Out[3]: array([[-2,  3],
               [ 4,  5]])

In [4]: scipy.linalg.det(a)
Out[4]: -22.0
```

The inverse of a matrix is computed using the `scipy.linalg.inv` function, while the product of two matrices is calculated using the NumPy `dot` function:

```
In [5]: b = scipy.linalg.inv(a)

In [6]: b
Out[6]: array([[-0.22727273,  0.13636364],
               [ 0.18181818,  0.09090909]])
```

```
In [7]: dot(a, b)
Out[7]: array([[ 1., 0.],
               [ 0., 1.]])
```

9.3.2 Solving systems of linear equations

Solving systems of equations is nearly as simple as constructing a coefficient matrix and a column vector. Suppose you have the following system of linear equations to solve:

$$\begin{align} 2x_1 + 4x_2 + 6x_3 &= 4 \\ x_1 - 3x_2 - 9x_3 &= -11 \\ 8x_1 + 5x_2 - 7x_3 &= 1 \end{align} \tag{9.1}$$

The first task is to recast this set of equations as a matrix equation of the form **Ax** = **b**. In this case, we have:

$$A = \begin{pmatrix} 2 & 4 & 6 \\ 1 & -3 & -9 \\ 8 & 5 & -7 \end{pmatrix}, \quad x = \begin{pmatrix} x_1 \\ x_2 \\ x_3 \end{pmatrix}, \quad b = \begin{pmatrix} 4 \\ -11 \\ 1 \end{pmatrix}. \tag{9.2}$$

Next we construct the NumPy arrays reprenting the matrix A and the vector **b**:

```
In [8]: A = array([[2, 4, 6], [1, -3, -9], [8, 5, -7]])
In [9]: b = array([4, -11, 2])
```

Finally we use the SciPy function `scipy.linalg.solve` to find x_1, x_2, and x_3:

```
In [10]: scipy.linalg.solve(A, b)
Out[10]: array([ -8.91304348, 10.2173913 , -3.17391304])
```

which gives the results: $x_1 = -8.91304348$, $x_2 = 10.2173913$, and $x_3 = -3.17391304$. Of course, you can get the same answer by noting that $x = A^{-1}b$. Following this approach, we can use *scipy.linalg.inv* introduced in the previous section:

```
In [11]: Ainv = scipy.linalg.inv(A)
```

```
In [12]: dot(Ainv, b)
Out[12]: array([ -8.91304348, 10.2173913 , -3.17391304])
```

which is the same answer we obtained using `scipy.linalg.solve`. Using `scipy.linalg.solve` is faster and numerically more stable than using $\mathbf{x} = A^{-1}\mathbf{b}$, so it is the preferred method for solving systems of equations.

You might wonder what happens if the system of equations are not all linearly independent. For example, if the matrix A is given by

$$A = \begin{pmatrix} 2 & 4 & 6 \\ 1 & -3 & -9 \\ 1 & 2 & 3 \end{pmatrix} \tag{9.3}$$

where the third row is a multiple of the first row. Let's try it out and see what happens. First we change the bottom row of the matrix A and then try to solve the system as we did before.

```
In [13]: A[2] = array([1, 2, 3])

In [14]: A
Out[14]: array([[ 2,  4,  6],
               [ 1, -3, -9],
               [ 1,  2,  3]])

In [15]: scipy.linalg.solve(A,b)
LinAlgError: Singular matrix

In [16]: Ainv = scipy.linalg.inv(A)
LinAlgError: Singular matrix
```

Whether we use `scipy.linalg.solve` or `scipy.linalg.inv`, SciPy raises an error because the matrix is singular.

9.3.3 Eigenvalue problems

One of the most common problems in science and engineering is the eigenvalue problem, which in matrix form is written as

$$A\mathbf{x} = \lambda \mathbf{x} \tag{9.4}$$

where A is a square matrix, $\mathbf{x}$ is a column vector, and λ is a scalar (number). Given the matrix A, the problem is to find the set of eigenvectors $\mathbf{x}$ and their corresponding eigenvalues λ that solve this equation.

We can solve eigenvalue equations like this using the SciPy routine `scipy.linalg.eig`. The output of this function is an array whose entries are the eigenvalues and a matrix whose rows are the eigenvectors. Let's return to the matrix we were using previously and find its eigenvalues and eigenvectors.

Numerical Routines: SciPy and NumPy

```
In [17]: A
Out[17]: array([[ 2,  4,  6],
                [ 1, -3, -9],
                [ 8,  5, -7]])

In [18]: lam, evec = scipy.linalg.eig(A)

In [19]: lam
Out[19]: array([ 2.40995356+0.j, -8.03416016+0.j,
          -2.37579340+0.j])

In [20]: evec
Out[20]: array([[-0.77167559, -0.52633654,  0.57513303],
                [ 0.50360249,  0.76565448, -0.80920669],
                [-0.38846018,  0.36978786,  0.12002724]])
```

The first eigenvalue and its corresponding eigenvector are given by

```
In [21]: lam[0]
Out[21]: (2.4099535647625494+0j)

In [22]: evec[:,0]
Out[22]: array([-0.77167559, 0.50360249, -0.38846018])
```

We can check that they satisfy the Ax = λx:

```
In [23]: dot(A,evec[:,0])
Out[23]: array([-1.85970234, 1.21365861,
          -0.93617101])

In [24]: lam[0]*evec[:,0]
Out[24]: array([-1.85970234+0.j, 1.21365861+0.j,
          -0.93617101+0.j])
```

Thus we see by direct substitution that the left and right sides of Ax = λx: are equal. In general, the eigenvalues can be complex, so their values are reported as complex numbers.

Generalized eigenvalue problem

The `scipy.linalg.eig` function can also solve the *generalized* eigenvalue problem

$$Ax = \lambda Bx \qquad (9.5)$$

where B is a square matrix with the same size as A. Suppose, for example, that we have

```
In [25]: A = array([[2, 4, 6], [1, -3, -9], [8, 5, -7]])
Out[25]: B = array([[5, 9, 1], [-3, 1, 6], [4, 2, 8]])
```

Then we can solve the generalized eigenvalue problem by entering B as the optional second argument to `scipy.linalg.eig`

```
In [26]: lam, evec = scipy.linalg.eig(A, B)
```

The solutions are returned in the same fashion as before, as an array `lam` whose entries are the eigenvalues and a matrix `evac` whose rows are the eigenvectors.

```
In [27]: lam
Out[27]: array([-1.36087907+0.j,  0.83252442+0.j,
               -0.10099858+0.j])

In [28]: evec
Out[28]: array([[-0.0419907 , -1.    ,  0.93037493],
                [-0.43028153,  0.17751302, -1.    ],
                [ 1.    , -0.29852465,  0.4226201 ]])
```

Hermitian and banded matrices

SciPy has a specialized routine for solving eigenvalue problems for Hermitian (or real symmetric) matrices. The routine for Hermitian matrices is `scipy.linalg.eigh`. It is more efficient (faster and uses less memory) than `scipy.linalg.eig`. The basic syntax of the two routines is the same, although some of the *optional* arguments are different. Both routines can solve generalized as well as standard eigenvalue problems.

SciPy has a specialized routine `scipy.linalg.eig_banded` for solving eigenvalue problems for real symmetric or complex Hermitian banded matrices. When there is a specialized routine for handling a particular kind of matrix, you should use it; it is almost certain to run faster, use less memory, and give more accurate results.

9.4 Solving Nonlinear Equations

SciPy has many different routines for numerically solving nonlinear equations or systems of nonlinear equations. Here we will introduce only a few of these routines, the ones that are relatively simple and appropriate for the most common types of nonlinear equations.

9.4.1 Single equations of a single variable

Solving a single nonlinear equation is enormously simpler than solving a system of nonlinear equations, so that is where we start. A word of caution: solving nonlinear equations can be a tricky business so it is important that you have a good sense of the behavior of the function you are trying to solve. The best way to do this is to plot the function over the domain of interest before trying to find the solutions. This will greatly assist you in finding the solutions you seek and avoiding spurious solutions.

We begin with a concrete example. Suppose we want to find the solutions to the equation

$$\tan x = \sqrt{(8/x)^2 - 1}. \tag{9.6}$$

Plots of $\tan x$ and $\sqrt{(8/x)^2 - 1}$ vs. x are shown in the top plot of Fig. 6.9, albeit with x replaced by θ. The solutions to this equation are those x values where the two curves $\tan x$ and $\sqrt{(8/x)^2 - 1}$ cross each other. The first step toward obtaining a numerical solution is to rewrite the equation to be solved in the form $f(x) = 0$. Doing so, the above equation becomes

$$\tan x - \sqrt{(8/x)^2 - 1} = 0. \tag{9.7}$$

Obviously the two equations above have the same solutions for x. Parenthetically we mention that the problem of finding the solutions to equations of the form $f(x) = 0$ is often referred to as *finding the roots* of $f(x)$.

Next, we plot $f(x)$ over the domain of interest, in this case from $x = 0$ to 8. For $x > 8$, the equation has no real solutions as the argument of the square root becomes negative. The solutions, points where $f(x) = 0$, are indicated by open green circles; there are three of them. Another notable feature of the function is that it diverges to $\pm\infty$ at $x = 0$, $\pi/2$, $3\pi/2$, and $5\pi/2$.

Brent method

One of the workhorses for finding solutions to a single variable nonlinear equation is the method of Brent, discussed in many texts on numerical methods. SciPy's implementation of the Brent algorithm is the function `scipy.optimize.brentq(f, a, b)`, which has three required

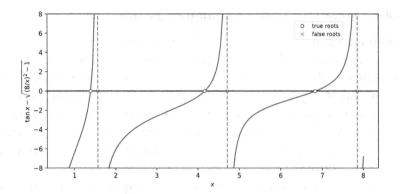

Figure 9.3 Roots of a nonlinear function.

arguments. The first argument f is the name of the user-defined function to be solved. The next two, a and b, are the x values that bracket the solution you are looking for. You should choose a and b so that there is only one solution in the interval between a and b. Brent's method also requires that f(a) and f(b) have opposite signs; an error message is returned if they do not. Thus to find the three solutions to $\tan x - \sqrt{(8/x)^2 - 1} = 0$, we need to run scipy.optimize.brentq(f, a, b) three times using three different values of a and b that bracket each of the three solutions. The program below illustrates the how to use scipy.optimize.brentq.

Code: chapter9/programs/rootbrentq.py

```
import numpy as np
import scipy.optimize
import matplotlib.pyplot as plt


def tdl(x):
    y = 8./x
    return np.tan(x) - np.sqrt(y*y-1.0)


# Find true roots
rx1 = scipy.optimize.brentq(tdl, 0.5, 0.49*np.pi)
rx2 = scipy.optimize.brentq(tdl, 0.51*np.pi, 1.49*np.pi)
rx3 = scipy.optimize.brentq(tdl, 1.51*np.pi, 2.49*np.pi)
rx = np.array([rx1, rx2, rx3])
ry = np.zeros(3)
# print true roots using a list comprehension
```

Numerical Routines: SciPy and NumPy

```
18  print('\nTrue roots:')
19  print('\n'.join('f({0:0.5f}) = {1:0.2e}'
20                 .format(x, tdl(x)) for x in rx))
21
22  # Find false roots
23  rx1f = scipy.optimize.brentq(tdl, 0.49*np.pi, 0.51*np.pi)
24  rx2f = scipy.optimize.brentq(tdl, 1.49*np.pi, 1.51*np.pi)
25  rx3f = scipy.optimize.brentq(tdl, 2.49*np.pi, 2.51*np.pi)
26  rxf = np.array([rx1f, rx2f, rx3f])
27  # print false roots using a list comprehension
28  print('\nFalse roots:')
29  print('\n'.join('f({0:0.5f}) = {1:0.2e}'
30                 .format(x, tdl(x)) for x in rxf))
31
32  # Plot function and various roots
33  x = np.linspace(0.7, 8, 128)
34  y = tdl(x)
35  # Create masked array for plotting
36  ymask = np.ma.masked_where(np.abs(y) > 20., y)
37
38  fig, ax = plt.subplots(figsize=(8, 4))
39  ax.plot(x, ymask)
40  ax.axhline(color='black')
41  ax.axvline(x=np.pi/2., color="gray",
42             linestyle='--', zorder=-1)
43  ax.axvline(x=3.*np.pi/2., color="gray",
44             linestyle='--', zorder=-1)
45  ax.axvline(x=5.*np.pi/2., color="gray",
46             linestyle='--', zorder=-1)
47  ax.set_xlabel(r'$x$')
48  ax.set_ylabel(r'$\tan\,x - \sqrt{(8/x)^2-1}$')
49  ax.set_ylim(-8, 8)
50
51  ax.plot(rx, ry, 'og', ms=5, mfc='white', label='true roots')
52
53  ax.plot(rxf, ry, 'xr', ms=5, label='false roots')
54  ax.legend(numpoints=1, fontsize='small',
55            loc='upper right',
56            bbox_to_anchor=(0.9, 0.97))
57  fig.tight_layout()
58  fig.savefig('figures/rootbrentq.pdf')
59  fig.show()
```

Running this code generates the following output:

```
In [1]: run rootbrentq.py

True roots:
f(1.39547) = -6.39e-14
f(4.16483) = -7.95e-14
```

```
f(6.83067) = -1.11e-15

False roots:
f(1.57080) = -1.61e+12
f(4.71239) = -1.56e+12
f(7.85398) = 1.17e+12
```

The Brent method finds the three true roots of the equation quickly and accurately when you provide values for the brackets a and b that are valid. However, like many numerical methods for finding roots, the Brent method can produce spurious roots as it does in the above example when a and b bracket singularities like those at $x = \pi/2$, $3\pi/2$, and $5\pi/2$. Here we evaluated the function at the purported roots found by brentq to verify that the values of x found were indeed roots. For the true roots, the values of the function were very near zero, to within an acceptable roundoff error of less than 10^{-13}. For the false roots, exceedingly large numbers on the order of 10^{12} were obtained, indicating a possible problem with these roots. These results, together with the plots, allow you to unambiguously identify the true solutions to this nonlinear function.

The brentq function has a number of optional keyword arguments that you may find useful. One keyword argument causes brentq to return not only the solution but the value of the function evaluated at the solution. Other arguments allow you to specify a tolerance to which the solution is found as well as a few other parameters possibly of interest. Most of the time, you can leave the keyword arguments at their default values. See the brentq entry online on the SciPy web site for more information.

Other methods for solving equations of a single variable

SciPy provides a number of other methods for solving nonlinear equations of a single variable. It has an implementation of the Newton-Raphson method called scipy.optimize.newton. It's the race car of such methods; it's super fast but less stable than the Brent method. To fully realize its speed, you need to specify not only the function to be solved, but also its first derivative, which is often more trouble than it's worth. You can also specify its second derivative, which may further speed up finding the solution. If you do not specify the first or second derivatives, the method uses the secant method, which is usually slower than the Brent method.

Other methods, including the Ridder (scipy.optimize.ridder)

ns and bisection (scipy.optimize.bisect) methods, are also available, although the Brent method is generally superior. SciPy lets you use your favorite.

9.4.2 Solving systems of nonlinear equations

Solving systems of nonlinear equations is not for the faint of heart. These are difficult problems that lack any general-purpose solutions. Nevertheless, SciPy provides quite an assortment of numerical solvers for nonlinear systems of equations. However, because of the complexity and subtleties of this class of problems, we do not discuss their use here.

9.5 Numerical Integration

When a function cannot be integrated analytically, or is very difficult to integrate analytically, one generally turns to numerical integration methods. SciPy has a number of routines for performing numerical integration. Most of them are found in the same `scipy.integrate` library where the ODE solvers are located. We list them in Table 9.2 for reference.

Function	Description
quad	single integration
dblquad	double integration
tplquad	triple integration
nquad	n-fold multiple integration
fixed_quad	Gaussian quadrature, order n
quadrature	Gaussian quadrature to tolerance
romberg	Romberg integration
trapz	trapezoidal rule
cumtrapz	trapezoidal rule to cumulatively compute integral
simps	Simpson's rule
romb	Romberg integration
polyint	Analytical polynomial integration (numpy)
poly1d	Helper function for polyint (numpy)

Table 9.2 Some integrating routines from `scipy.integrate` unless otherwise noted.

9.5.1 Single integrals

The function `quad` is the workhorse of SciPy's integration functions. Numerical integration is sometimes called *quadrature*, hence the name. The function `quad` is the default choice for performing single integrals of a function $f(x)$ over a given fixed range from a to b

$$\int_a^b f(x)\,dx. \tag{9.8}$$

The general form of `quad` is `scipy.integrate.quad(f, a, b)`, where `f` is the *name* of the function to be integrated and `a` and `b` are the lower and upper limits, respectively. The routine uses *adaptive quadrature* methods to numerically evaluate integrals, meaning it successively refines the subintervals (makes them smaller) until a desired level of numerical precision is achieved. For the `quad` routine, this is about 10^{-8}, although it usually does even better.

As an example, let's integrate a Gaussian function over the range from 0 to 1:

$$\int_0^1 e^{-x^2}\,dx \tag{9.9}$$

We first need to define the function $f(x) = e^{-x^2}$, which we do using a lambda expression, and then we call the function `quad` to perform the integration.

```
In [1]: f = lambda x : exp(-x**2)

In [2]: from scipy.integrate import quad

In [3]: quad(f, 0, 1)
Out[3]: (0.7468241328124271, 8.291413475940725e-15)
```

The function call `scipy.integrate.quad(f, 0, 1)` returns two numbers. The first is `0.7468...`, which is the value of the integral, and the second is `8.29...e-15`, which is an estimate of the absolute error in the value of the integral, which we see is quite small compared to `0.7468`.

For its first argument, `quad` requires a function *name*. In this case, we used a lambda expression to define the function name, `f` in this case. Alternatively, we could have defined the function using the usual `def` construction:

```
def f(x):
    return np.exp(-x**2)
```

Numerical Routines: SciPy and NumPy

But here it was simpler to use a lambda expression. Even simpler, we can just put the lambda expression directly into the first argument of `quad`, as illustrated here:

```
In [4]: quad(lambda x : exp(-x**2), 0, 1)
Out[4]: (0.7468241328124271, 8.291413475940725e-15)
```

That works too! Thus we see a `lambda` expression used as an *anonymous function*, a function with no name, as promised in §7.3.

Interestingly, the `quad` function accepts positive and negative infinity as limits.

```
In [5]: quad(lambda x : exp(-x**2), 0, inf)
Out[5]: (0.8862269254527579, 7.101318390472462e-09)

In [6]: scipy.integrate.quad(lambda x : exp(-x**2), -inf, 1)
Out[6]: (1.6330510582651852, 3.669607414547701e-11)
```

The `quad` function handles infinite limits just fine. The absolute errors are somewhat larger but still well within acceptable bounds for practical work. Note that `inf` is a NumPy object and should be written as `np.inf` within a Python program.

The `quad` function can integrate standard predefined NumPy functions of a single variable, like `exp`, `sin`, and `cos`.

```
In [7]: quad(exp, 0, 1)
Out[7]: (1.7182818284590453, 1.9076760487502457e-14)

In [8]: quad(sin, -0.5, 0.5)
Out[8]: (0.0, 2.707864644566304e-15)

In [9]: quad(cos, -0.5, 0.5)
Out[9]: (0.9588510772084061, 1.0645385431034061e-14)
```

Suppose we want to integrate a function such as Ae^{-cx^2} defined as a normal Python function:

```
In [10]: def gauss(x, A, c):
   ....:     return A * np.exp(-c*x**2)
```

Of course we will need to pass the values of A and c to `gauss` via `quad` in order to numerically perform the integral. This can be done using `args`, one of the optional keyword arguments of `quad`. The code below shows how to do this

```
In [11]: A, c = 2.0, 0.5

In [12]: intgrl1 = quad(gauss, 0.0, 5.0, args=(A, c))
```

```
In [13]: intgrl1
Out[13]: (2.5066268375731307, 2.1728257867977207e-09)
```

Note that the order of the additional parameters in `args=(A, c)` must be in the same order as they appear in the function definition of `gauss`.

Of course, we could also do this using a lambda expression, as follows

```
In [14]: intgrl2 = quad(lambda x: gauss(x, A, c), 0.0, 5.0)
```

```
In [15]: intgrl2
Out[15]: (2.5066268375731307, 2.1728257867977207e-09)
```

Either way, we get the same answer.

Let's do one last example. Let's integrate the first-order Bessel function of the first kind, usually denoted $J_1(x)$, over the interval from 0 to 5. $J_1(x)$ is available in the special functions library of SciPy as `scipy.special.jn(v, x)` where v is the (real) order of the Bessel function (see §9.1). Note that x is the *second* argument of `scipy.special.jn(v, x)`, which means that we cannot use the args keyword function because the integration routine quad *assumes* that the independent variable is the first argument of the function to be integrated. Here the first argument is v, which we wish to fix to be 1. Therefore, we use a lambda expression, to fix the parameters A and c, assign 1 to the value of v, and to define the function to be integrated. Here is now it works:

```
In [10]: import scipy.special
```

```
In [11]: quad(lambda x: scipy.special.jn(1,x), 0, 5)
Out[11]: (1.177596771314338, 1.8083362065765924e-14)
```

Because the SciPy function `scipy.special.jn(v, x)` is a function of two variables, v and x, and we want to use the second variable x as the independent variable, we cannot use the function name `scipy.special.jn` together with the args argument of quad. So we use a lambda expression, which is a function of only one variable, x, and set the v argument equal to 1.

Integrating polynomials

Working in concert with the NumPy `poly1d`, the NumPy function `polyint` takes the nth antiderivative of a polynomial and can be used to evaluate definite integrals. The function `poly1d` essentially does the same thing as `polyval`, which we encountered in §9.1, but with a different syntax. Suppose we want to make the polynomial function $p(x) = 2x^2 + 5x + 1$. Then we write

```
In [12]: p = np.poly1d([2, 5, 1])

In [13]: p
Out[13]: poly1d([2, 5, 1])
```

The polynomial $p(x) = 2x^2 + 5x + 1$ is evaluated using the syntax `p(x)`. Below, we evaluate the polynomial at three different values of x. The polynomial $p(x) = 2x^2 + 5x + 1$ is evaluated using the syntax `p(x)`. Below, we evaluate the polynomial at three different values of x.

```
In [14]: p(1), p(2), p(3.5)
Out[14]: p(8, 19, 43.0)
```

Thus `polyval` allows us to define the function $p(x) = 2x^2 + 5x + 1$. Now the antiderivative of $p(x) = 2x^2 + 5x + 1$ is $P(x) = \frac{2}{3}x^3 + \frac{5}{2}x^2 + x + C$ where C is the integration constant. The NumPy function `polyint`, which takes the nth antiderivative of a polynomial, works as follows

```
In [15]: P = polyint(p)

In [16]: P
Out[16]: poly1d([ 0.66666667, 2.5       , 1.       , 0.       ])
```

When `polyint` has a single input, p in this case, `polyint` returns the coefficients of the antiderivative with the integration constant set to zero, as Out[16] illustrates. It is then an easy matter to determine any definite integral of the polynomial $p(x) = 2x^2 + 5x + 1$ since

$$q \equiv \int_a^b p(x)\,dx = P(b) - P(a). \tag{9.10}$$

For example, if $a = 1$ and $b = 5$,

```
In [17]: q=P(5)-P(1)

In [18]: q
Out[18]: 146.66666666666666
```

or

$$\int_1^5 \left(2x^2 + 5x + 1\right)dx = 146\tfrac{2}{3}. \tag{9.11}$$

9.5.2 Double integrals

The `scipy.integrate` function `dblquad` can be used to numerically evaluate double integrals of the form

$$\int_{y=a}^{y=b} dy \int_{x=g(y)}^{x=h(y)} dx\, f(x,y). \qquad (9.12)$$

The general form of `dblquad` is

```
In [19]: scipy.integrate.dblquad(func, a, b, gfun, hfun)
```

where `func` is the name of the function to be integrated, `a` and `b` are the lower and upper limits of the y variable, respectively, and `gfun` and `hfun` are the *names* of the functions that define the lower and upper limits of the x variable. As an example, let's perform the double integral

$$\int_0^{1/2} dy \int_0^{\sqrt{1-4y^2}} 16xy\, dx. \qquad (9.13)$$

We define the functions f, g, and h, using lambda expressions. Note that even if g and h are constants, as they may be in many cases, they must be defined as functions, as we have done here for the lower limit.

```
In [20]: f = lambda x, y : 16*x*y

In [21]: g = lambda x : 0

In [22]: h = lambda y : sqrt(1-4*y**2)

In [23]: scipy.integrate.dblquad(f, 0, 0.5, g, h)
Out[23]: (0.5, 5.551115123125783e-15)
```

Once again, there are two outputs: the first is the value of the integral and the second is its absolute uncertainty.

Of course, the lower limit can also be a function of y, as we demonstrate here by performing the integral

$$\int_0^{1/2} dy \int_{1-2y}^{\sqrt{1-4y^2}} 16xy\, dx. \qquad (9.14)$$

The code for this is given by

```
In [24]: g = lambda y : 1-2*y

In [25]: scipy.integrate.dblquad(f, 0, 0.5, g, h)
Out[25]: (0.33333333333333326, 3.700743415417188e-15)
```

Numerical Routines: SciPy and NumPy

Other integration routines

In addition to the routines described above, `scipy.integrate` has a number of other integration routines, including `nquad`, which performs n-fold multiple integration, as well as other routines that implement other integration algorithms. You will find, however, that `quad` and `dblquad` meet most of your needs for numerical integration.

9.6 Solving ODEs

The `scipy.integrate` library has two powerful routines, `ode` and `odeint`, for numerically solving systems of coupled first-order ordinary differential equations (ODEs). While `ode` is more versatile, `odeint` (ODE integrator) has a simpler Python interface that works very well for most problems. It can handle both stiff and non-stiff problems. Here we provide an introduction to `odeint`.

A typical problem is to solve a second- or higher-order ODE for a given set of initial conditions. Here we illustrate using `odeint` to solve the equation for a driven damped pendulum. The equation of motion for the angle θ that the pendulum makes with the vertical is given by

$$\frac{d^2\theta}{dt^2} = \frac{1}{Q}\frac{d\theta}{dt} + \sin\theta + d\cos\Omega t \tag{9.15}$$

where t is time, Q is the quality factor, d is the forcing amplitude, and Ω is the driving frequency of the forcing. Reduced variables have been used such that the natural (angular) frequency of oscillation is 1. The ODE is nonlinear owing to the $\sin\theta$ term. Of course, it's precisely because there are no general methods for solving nonlinear ODEs that one employs numerical techniques, so it seems appropriate that we illustrate the method with a nonlinear ODE.

The first step is always to transform any n^{th}-order ODE into a system of n first-order ODEs of the form:

$$\begin{aligned}\frac{dy_1}{dt} &= f_1(t, y_1, ..., y_n) \\ \frac{dy_2}{dt} &= f_2(t, y_1, ..., y_n) \\ \vdots &= \vdots \\ \frac{dy_n}{dt} &= f_n(t, y_1, ..., y_n).\end{aligned} \tag{9.16}$$

We also need n initial conditions, one for each variable y_i. Here we have a second-order ODE so we will have two coupled ODEs and two initial conditions.

We start by transforming our second-order ODE into two coupled first-order ODEs. The transformation is easily accomplished by defining a new variable $\omega \equiv d\theta/dt$. With this definition, we can rewrite our second-order ODE as two coupled first-order ODEs:

$$\frac{d\theta}{dt} = \omega \tag{9.17}$$

$$\frac{d\omega}{dt} = \frac{1}{Q}\omega + \sin\theta + \cos\Omega t. \tag{9.18}$$

In this case the functions on the right-hand side of the equations are

$$f_1(t,\theta,\omega) = \omega \tag{9.19}$$

$$f_2(t,\theta,\omega) = -\frac{1}{Q}\omega + \sin\theta + d\cos\Omega t. \tag{9.20}$$

Note that there are no explicit derivatives on the right-hand side of the functions f_i; they are all functions of t and the various y_i, in this case θ and ω.

The initial conditions specify the values of θ and ω at $t = 0$.

SciPy's ODE solver `scipy.integrate.odeint` has three required arguments and many optional keyword arguments, of which we only need one, `args`, for this example. So in this case, `odeint` has the form

`odeint(func, y0, t, args=())`

The first argument `func` is the name of a Python function that returns a list of values of the n functions $f_i(t, y_1, ..., y_n)$ at a given time t. The second argument `y0` is an array (or list) of the values of the initial conditions of $(y_1, ..., y_n)$. The third argument is the array of times at which you want `odeint` to return the values of $(y_1, ..., y_n)$. The keyword argument `args` is a tuple that is used to pass parameters (besides `y0` and `t`) that are needed to evaluate `func`. Our example should make all of this clear.

Having written the nth-order ODE as a system of n first-order ODEs, the next task is to write the function `func`. The function `func` should have three arguments: (1) the list (or array) of current y values, (2) the current time `t`, and (3) a list of any other parameters `params` needed to evaluate `func`. The function `func` returns the values of the

Numerical Routines: SciPy and NumPy

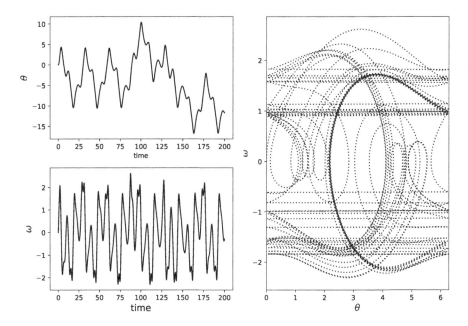

Figure 9.4 Pendulum trajectory.

derivatives $dy_i/dt = f_i(t, y_1, ..., y_n)$ in a list (or array). Lines 6–11 illustrate how to write func for our example of a driven damped pendulum. Here we name the function simply f, which is the name that appears in the call to odeint in line 35.

The only other tasks remaining are to define the parameters needed in the function, bundle them into a list (line 24), define the initial conditions, and bundle them into another list (line 27). After defining the time array in lines 30–32, we call odeint with the appropriate arguments and a variable, psoln in this case, to store output. The output psoln is an n element array where each element is itself an array corresponding to the values of y_i for each time in the time t array that was an argument of odeint. For this example, the first element psoln[:,0] is the y_0 or theta array, and the second element psoln[:,1] is the y_1 or omega array. The remainder of the code simply plots out the results in different formats. The resulting plots are shown in Fig. 9.4.

Code: chapter9/programs/odePend.py

```
1  import numpy as np
2  import matplotlib.pyplot as plt
```

```python
from scipy.integrate import odeint


def f(y, t, params):
    theta, omega = y      # unpack current values
    Q, d, Omega = params  # unpack parameters
    derivs = [omega,      # list of dy/dt=f functions
             -omega/Q+np.sin(theta)+d*np.cos(Omega*t)]
    return derivs


# Parameters
Q = 2.0            # quality factor (inverse damping)
d = 1.5            # forcing amplitude
Omega = 0.65       # drive frequency

# Initial values
theta0 = 0.0       # initial angular displacement
omega0 = 0.0       # initial angular velocity

# Bundle parameters for ODE solver
params = [Q, d, Omega]

# Bundle initial conditions for ODE solver
y0 = [theta0, omega0]

# Make time array for solution
tStop = 200.
tInc = 0.05
t = np.arange(0., tStop, tInc)

# Call the ODE solver
psoln = odeint(f, y0, t, args=(params,))

# Plot results
fig = plt.figure(figsize=(9.5, 6.5))

# Plot theta as a function of time
ax1 = fig.add_subplot(221)
ax1.plot(t, psoln[:, 0], color='black')
ax1.set_xlabel('time')
ax1.set_ylabel(r'$\theta$', fontsize=14)

# Plot omega as a function of time
ax2 = fig.add_subplot(223)
ax2.plot(t, psoln[:, 1], color='black')
ax2.set_xlabel('time', fontsize=14)
ax2.set_ylabel(r'$\omega$', fontsize=14)
```

```
52  # Plot omega vs theta
53  ax3 = fig.add_subplot(122)
54  twopi = 2.0*np.pi
55  ax3.plot(psoln[:, 0] % twopi, psoln[:, 1],
56          dashes=(1, 2), ms=1, color='black')
57  ax3.set_xlabel(r'$\theta$', fontsize=14)
58  ax3.set_ylabel(r'$\omega$', fontsize=14)
59  ax3.set_xlim(0., twopi)
60
61  fig.tight_layout()
62  fig.savefig('figures/odePend.pdf')
63  fig.show()
```

The plots in Fig. 9.4 reveal that for the particular set of input parameters chosen, `Q = 2.0, d = 1.5,` and `Omega = 0.65`, the pendulum trajectories are chaotic. Weaker forcing (smaller *d*) leads to what is perhaps the more familiar behavior of sinusoidal oscillations with a fixed frequency which, at long times, is equal to the driving frequency.

9.7 Discrete (Fast) Fourier Transforms

The SciPy library has a number of routines for performing discrete Fourier transforms. Before delving into them, we provide a brief review of Fourier transforms and discrete Fourier transforms.

9.7.1 Continuous and discrete Fourier transforms

The Fourier transform of a function $g(t)$ is given by

$$G(f) = \int_{-\infty}^{\infty} g(t)e^{-i2\pi ft} dt , \qquad (9.21)$$

where f is the Fourier transform variable; if t is time, then f is frequency. The inverse transform is given by

$$g(t) = \int_{-\infty}^{\infty} G(f)e^{i2\pi ft} dt . \qquad (9.22)$$

Here we define the Fourier transform in terms of the frequency f rather than the angular frequency $\omega = 2\pi f$.

The conventional Fourier transform is defined for continuous functions, or at least for functions that are dense and thus have an infinite number of data points. When doing numerical analysis, however, you work with *discrete* data sets, that is, data sets defined for a finite number of points. The discrete Fourier transform (DFT) is defined for a function g_n consisting of a set of N discrete data points. Those N data points must be defined at *equally spaced* times $t_n = n\Delta t$ where Δt is the time between successive data points and n runs from 0 to $N - 1$. The discrete Fourier transform (DFT) of g_n is defined as

$$G_l = \sum_{n=0}^{N-1} g_n e^{-i(2\pi/N)ln} \qquad (9.23)$$

where l runs from 0 to $N - 1$. The inverse discrete Fourier transform (iDFT) is defined as

$$g_n = \frac{1}{N} \sum_{l=0}^{N-1} G_l e^{i(2\pi/N)ln}. \qquad (9.24)$$

The DFT is usually implemented on computers using the well-known Fast Fourier Transform (FFT) algorithm, generally credited to Cooley and Tukey who developed it at AT&T Bell Laboratories during the 1960s. But their algorithm is essentially one of many independent rediscoveries of the basic algorithm dating back to Gauss who described it as early as 1805.

9.7.2 The SciPy FFT library

The SciPy library `scipy.fftpack` has routines that implement a souped-up version of the FFT algorithm along with many ancillary routines that support working with DFTs. The basic FFT routine in `scipy.fftpack` is appropriately named `fft`. The program below illustrates its use, along with the plots that follow.

Code: chapter9/programs/fftExample.py

```
1  import numpy as np
2  from scipy import fftpack
3  import matplotlib.pyplot as plt
4
5  width = 2.0
6  freq = 0.5
7
8  t = np.linspace(-10, 10, 128)
9  g = np.exp(-np.abs(t)/width) * np.sin(2.0*np.pi*freq*t)
10 dt = t[1]-t[0]   # increment between times in time array
11
```

Numerical Routines: SciPy and NumPy

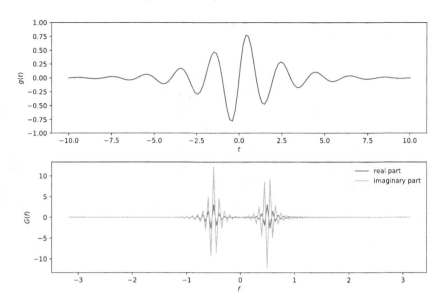

Figure 9.5 Function $g(t)$ and its DFT $G(f)$.

```
12  G = fftpack.fft(g)        # FFT of g
13  f = fftpack.fftfreq(g.size, d=dt)   # FFT frequenies
14  f = fftpack.fftshift(f)   # shift freqs from min to max
15  G = fftpack.fftshift(G)   # shift G order to match f
16
17  fig, (ax1, ax2) = plt.subplots(2, 1, figsize=(9, 6))
18  ax1.plot(t, g)
19  ax1.set_xlabel(r'$t$')
20  ax1.set_ylabel(r'$g(t)$')
21  ax1.set_ylim(-1, 1)
22  ax2.plot(f, np.real(G), color='dodgerblue',
23          label='real part')
24  ax2.plot(f, np.imag(G), color='coral',
25          label='imaginary part')
26  ax2.legend()
27  ax2.set_xlabel(r'$f$')
28  ax2.set_ylabel(r'$G(f)$')
29  plt.tight_layout()
30  plt.savefig('figures/fftExample.pdf')
31  plt.show()
```

The DFT has real and imaginary parts, both of which are plotted in Fig. 9.5.

The fft function returns the N Fourier components of G_n starting with the zero-frequency component G_0 and progressing to the maximum positive frequency component $G_{(N/2)-1}$ (or $G_{(N-1)/2}$ if N is odd). From there, fft returns the maximum *negative* component $G_{N/2}$ (or $G_{(N-1)/2}$ if N is odd) and continues upward in frequency until it reaches the minimum negative frequency component G_{N-1}. This is the standard way that DFTs are ordered by most numerical DFT packages. The scipy.fftpack function fftfreq creates the array of frequencies in this non-intuitive order such that f[n] in the above routine is the correct frequency for the Fourier component G[n]. The arguments of fftfreq are the size of the original array g and the keyword argument d that is the spacing between the (equally spaced) elements of the time array (d=1 if left unspecified). The package scipy.fftpack provides the convenience function fftshift that reorders the frequency array so that the zero-frequency occurs at the middle of the array, that is, so the frequencies proceed monotonically from smallest (most negative) to largest (most positive). Applying fftshift to both f and G puts the frequencies f in ascending order and shifts G so that the frequency of G[n] is given by the shifted f[n].

The scipy.fftpack module also contains routines for performing 2-dimensional and n-dimensional DFTs, named fft2 and fftn, respectively, using the FFT algorithm.

As for most FFT routines, the scipy.fftpack FFT routines are most efficient if N is a power of 2. Nevertheless, the FFT routines are able to handle data sets where N is not a power of 2.

scipy.fftpack also supplies an inverse DFT function ifft. It is written to act on the *unshifted* FFT so take care! Note also that ifft returns a *complex* array. Because of machine roundoff error, the imaginary part of the function returned by ifft will, in general, be very near zero but not exactly zero even when the original function is a purely real function.

9.8 Exercises

1. Numerically solve the following system of equations

$$\begin{aligned} x_1 - 2x_2 + 9x_3 + 13x_4 &= 1 \\ -5x_1 + x_2 + 6x_3 - 7x_4 &= -3 \\ 4x_1 + 8x_2 - 4x_3 - 2x_4 &= -2 \\ 8x_1 + 5x_2 - 7x_3 + x_4 &= 5 \end{aligned} \quad (9.25)$$

2. Numerically integrate the following integrals and compare the results to the exact value of each one.

$$\int_{-1}^{1} \frac{dx}{1+x^2} = \frac{\pi}{2} \qquad \int_{-\infty}^{\infty} \frac{dx}{(e^x + x + 1)^2 + \pi^2} = \frac{2}{3}$$

3. Use `scipy.integrate.odeint` to solve the following set of nonlinear ODEs.

$$\frac{dx}{dt} = a(y-x), \qquad \frac{dy}{dt} = (c-a)x - xz + cy, \qquad \frac{dz}{dt} = xy - bz$$

For the initial conditions, use $x_0 = -10$, $y_0 = 0$, $z_0 = 35$. Setting the initial parameters to $a = 40$, $b = 5$, $c = 35$ gives chaotic solutions like those shown below. Setting $b = 10$ while keeping $a = 40$ and $c = 35$ yields periodic solutions. Take care to choose a small enough time step (but not too small!).

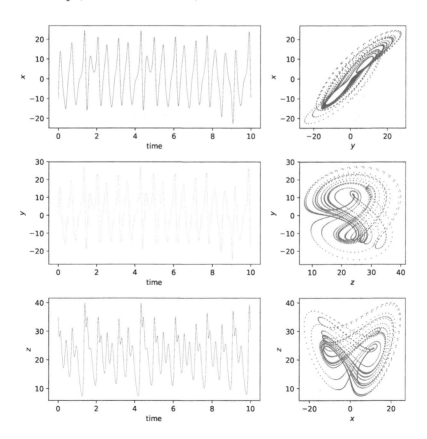

4. In this exercise, you explore the use of discrete Fourier transforms to filter noisy signals. As a first step, use the following function to create a noisy Gaussian waveform:

```python
def gaussNoisy(x, noiseAmp):
    noise = noiseAmp*(np.random.randn(len(x)))
    return np.exp(-0.5*x*x) * (1.0+noise)
N = 256
x = np.linspace(-4.0, 4.0, N)
y = gaussNoisy(x, 0.1)
```

(a) Calculate the discrete Fourier transform using NumPy's `fft` and `fftshift` routines so that you have a properly ordered Fourier transform.

(b) Plot the noisy Gaussian and its DFT on two separate panes in the same figure window. Set the limits of the y-axis of the DFT plot so that you can see the noise at the high frequencies (limits of ±2 should suffice).

(c) Next, set all the frequencies higher than a certain cutoff equal to zero (real and imaginary parts) to obtain a filtered DFT (take care to do this right!). Inverse Fourier transform the filtered DFT. Then, add a third frame to your figure and plot the inverse transform of the filtered DFT as well as the original. If it works well, you should observe a smooth Gaussian that goes through the original noisy one. Experiment with different cutoffs, say, all the frequencies above 2, 4, 8, 16, 32. The figure below shows an example of what the third frame of your plot might look like.

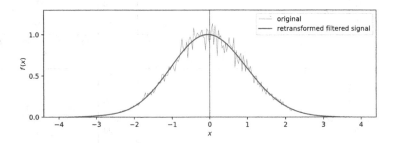

5. Use NumPy's `polyval` function together with SciPy or SciPy's `special.eval_chebyt` function to plot the following functions:

(a) The first four Chebyshev polynomials of the first kind over the interval from -1 to $+1$. Look up `scipy.special.chebyt` on the SciPy web site.

(b) The first four Hermite polynomials *multiplied* by

$$\frac{e^{-x^2/2}}{\left(2^n n! \sqrt{\pi}\right)^{1/2}}. \tag{9.26}$$

Plot these on the interval from -5 to $+5$. They are the first four wave functions of the quantum mechanical simple harmonic oscillator.

CHAPTER 10

Data Manipulation and Analysis: Pandas

> *This chapter introduces Pandas, a powerful Python package for manipulating and analyzing large (and small) data sets. You first learn how to **read data** from external files, e.g., Excel or text files, into Pandas. You learn about different **data structures** for storing dates and times, **time series**, and data organized into rows and columns in a spreadsheet-like structure called a **DataFrame**. You then learn how to manipulate data, how to extract subsets of data, and how to plot those data using matplotlib, but with some new syntax introduced by Pandas that facilitates working with the data structures of Pandas.*

In Chapter 4 we introduced a few simple methods for reading and writing data from and to data files using some routines available in NumPy. In this chapter, we introduce more versatile and powerful methods for reading, writing, and more importantly, *manipulating* large (and small) data sets using a Python package called *Pandas*.

Pandas is a versatile package for handling time series and large data sets. It has a spreadsheet-like character and was developed primarily for people working in the financial industry, with its own ecosystem that reflects its origins. Nevertheless, many of its features are generally useful to scientists and engineers working across a broad range of applications. We can't cover all of its capabilities in one short chapter, but we will attempt to show you a few of the things it can do. With that introduction, our hope is that you will have learned enough to adapt Pandas to your own applications.

Pandas comes installed with the standard Python distributions, like Enthought and Anaconda. In an IPython shell or in a Python program, you access the many routines available in Pandas by writing:

```
import pandas as pd
```

The abbreviation universally used for Pandas is `pd`.

10.1 Reading Data from Files Using Pandas

Pandas can read data from files written in many different formats, including the following: text, csv, Excel, JSON (JavaScript Object Notation), fixed-width text tables, HTML (web pages), and more that you can define. Our purpose here, however, is not to exhaust all the possibilities. Instead, we show you some of the more common methods and also show you a few tricks. The idea is to illustrate, with some well-chosen examples (we hope!), how you can use Pandas, so that you get the idea of how Pandas works. When we are finished, you should be able to use Pandas to read in and manipulate data, and then also be able to read the appropriate online Pandas documentation to extend your knowledge and adapt Pandas for your own applications.

10.1.1 Reading from Excel files saved as csv files

Excel files are commonly used to store data. As we learned in Chapter 4, one simple way to read in data from an Excel file is to save it as a csv file, which is a text file of tabular data with different columns of data separated by commas (hence the name csv: comma-separated values, see §4.3.2).

Let's start with the Excel file shown in Fig. 10.1. The Excel application can save the spreadsheet as a csv text file. It looks like this:

Data: chapter10/programs/ScatMieData.csv

```
Wavelength [vacuum] (nm) = 532,,
Refractive index of solvent = 1.33,,
Refractive index of particles = 1.59,,
Diameter of particles (microns) = 0.5,,
Cos_theta,F1,F2
1.00E+00,7.00E+01,7.00E+01
8.75E-01,2.71E+01,2.35E+01
7.50E-01,8.58E+00,6.80E+00
6.25E-01,1.87E+00,1.72E+00
5.00E-01,2.25E-01,5.21E-01
3.75E-01,3.04E-01,3.11E-01
2.50E-01,6.54E-01,2.36E-01
1.25E-01,7.98E-01,1.49E-01
0.00E+00,7.04E-01,7.63E-02
-1.25E-01,4.85E-01,4.06E-02
-2.50E-01,2.65E-01,3.64E-02
-3.75E-01,1.17E-01,4.59E-02
-5.00E-01,6.23E-02,5.79E-02
-6.25E-01,8.51E-02,7.63E-02
-7.50E-01,1.56E-01,1.20E-01
```

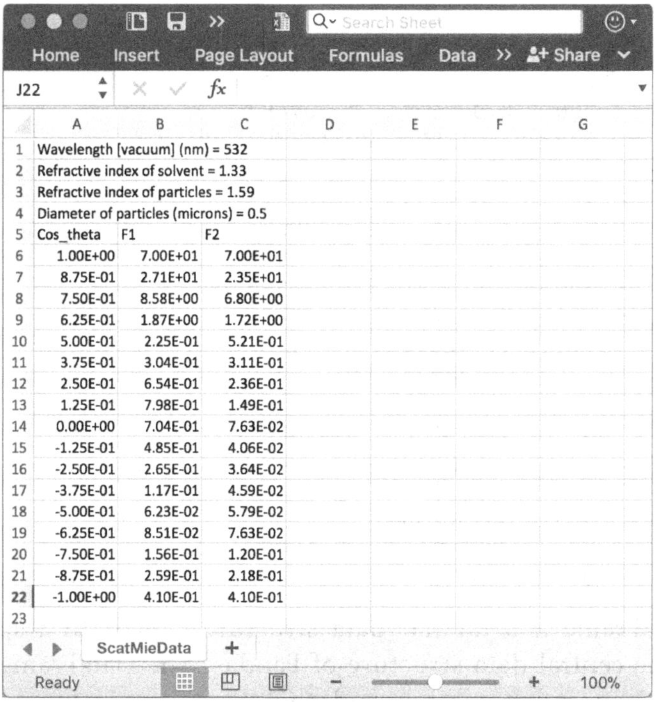

Data Manipulation and Analysis: Pandas 241

Figure 10.1   Excel spreadsheet.

```
-8.75E-01,2.59E-01,2.18E-01
-1.00E+00,4.10E-01,4.10E-01
```

This particular file has a header that provides information about the data, a header row specifying the a name of each column of data, Cos_theta,[1] F1, and F2, followed by three columns of data.

Let's use Pandas to read in the data in this file. To start, we skip the header information contained in the top 4 lines of the file using the skiprows keyword argument:

```
In [1]: scat = pd.read_csv('ScatMieData.csv', skiprows=4)
```

The Pandas function pd.read_csv() reads the data into a special Pandas object called a *DataFrame*, to which we give the name scat in the code above. We can examine the DataFrame by typing scat at the IPython prompt:

---

[1] In general, it's preferable to use column names that have no spaces, which is why we have used an underline here. Pandas can handle headers with spaces, although in some cases it can be limiting.

```
In [2]: scat
Out[2]:
    Cos_theta          F1          F2
0       1.000   70.007687   70.007687
1       0.875   27.111508   23.482798
2       0.750    8.577917    6.797595
3       0.625    1.866338    1.722973
4       0.500    0.224671    0.520981
5       0.375    0.304370    0.310918
6       0.250    0.653748    0.235945
7       0.125    0.798098    0.149235
8       0.000    0.703536    0.076290
9      -0.125    0.485046    0.040556
10     -0.250    0.264527    0.036360
11     -0.375    0.116618    0.045887
12     -0.500    0.062257    0.057888
13     -0.625    0.085123    0.076260
14     -0.750    0.156296    0.119770
15     -0.875    0.259113    0.218251
16     -1.000    0.410416    0.410416
```

A DataFrame is a tabular data structure similar to a spreadsheet. It is the central data structure of Pandas. The DataFrame `scat` consists of an index column and the three data columns from the `ScatMieData.csv` data file. The index column is added by Pandas and runs from 0 to $N-1$, where N is the number of data points in the file. The three data columns are labeled with the names given in the fifth line of the data file, which was the first line read by `pd.read_csv()`, as the keyword argument `skiprows` was set equal to 4. By default, Pandas assumes that the first line read gives the names of the data columns that follow.

The data in the DataFrame can be accessed, and sliced and diced, in different ways. To access the data in the column, you use the column labels:

```
In [3]: scat.Cos_theta
Out[3]:
0     1.000
1     0.875
2     0.750
3     0.625
4     0.500
5     0.375
6     0.250
7     0.125
8     0.000
9    -0.125
```

Data Manipulation and Analysis: Pandas

```
10    -0.250
11    -0.375
12    -0.500
13    -0.625
14    -0.750
15    -0.875
16    -1.000
Name: Cos_theta, dtype: float64
```

Typing `scat['Cos_theta']`, a syntax similar to the one used for dictionaries, yields the same result. Individual elements and slices can be accessed by indexing as for NumPy arrays:

```
In [4]: scat.Cos_theta[2]
Out[4]: 0.75

In [5]: scat.Cos_theta[2:5]
Out[5]:
2    0.750
3    0.625
4    0.500
Name: Cos_theta, dtype: float64

In [6]: scat['Cos_theta'][2]
Out[6]: 0.75
```

Similarly, `scat.F1` and `scat.F2` give the data in the columns labeled F1 and F2. We will return to the subject of DataFrames in §10.3.2.

In the example above, we ignored the header data contained in the first 4 lines by setting `skiprows=4`. But suppose we want to read in the information in those four rows. How would we do it? Let's try the Panda routine `read_csv()` once again.

```
In [7]: head = pd.read_csv('ScatMieData.csv', nrows=4,
   ...:                    header=None)
```

We use the keyword `nrows` and set it equal to 4 so that Pandas reads only the first 4 lines of the file, which comprise the header information. We also set `head=None` as there is no separate header information for these 4 rows. We examine the result by typing `head`, the name we assigned to these data.

```
In [8]: head
Out[8]:
                                         0       1    2
0          Wavelength [vacuum] (nm) = 532     NaN  NaN
1         Refractive index of solvent = 1.33   NaN  NaN
2       Refractive index of particles = 1.59   NaN  NaN
3      Diameter of particles (microns) = 0.5   NaN  NaN
```

The 4 rows are indexed from 0 to 3, as expected. The 3 columns are indexed from 0 to 2. Pandas introduced the column indices 0 to 2 because we set `header=None` in the `read_csv()` calling function, instead of inferring names for these columns from the first row read from the csv file as it did above. Individual elements of the `head` DataFrame can be indexed by their column and row, respectively:

```
In [9]: head[0][1]
Out[9]: 'Refractive index of solvent = 1.33'
```

The quotes tell us that the output of `head[0][1]` is a string. In general, Pandas infers the data types for the different columns and assigns them correctly for numeric, string, or Boolean data types. But in this case, the information in the first column of the spreadsheet contains both string and numeric data, so `read_csv()` interprets the column as a string.

The data in the other two columns, which were empty in the original Excel spreadsheet and are commas with nothing between them in the csv file, become `NaN` ("not a number") in the DataFrame. Accessing the datum in one cell as before gives the expected result:

```
In [10]: head[1][1]
Out[10]: nan
```

Pandas fills in missing data with `NaN`, a feature we discuss in greater detail in §10.3.2.

The appearance of quotes in the output `Out[8]:` above indicates that the data read from the header information are stored as strings. We might prefer, however, to separate out the numeric data from the strings that describe them. While Python has routines for stripping off numeric information from strings, it's more efficient to perform this task when reading the file. To do this, we use the Pandas routine `read_table()`, which, like `read_csv()`, reads in data from a text file. With `read_table()`, however, the user can specify the symbol that will be used to separate the columns of data: a symbol other than a comma can be used. The following call to `read_table()` does just that.

```
In [11]: head = pd.read_table('ScatMieData.csv', sep='=',
   ...:    nrows=4, header=None)
```

The keyword `sep`, which specifies the symbol that separates columns, is set equal to the string `'='`, as the equals sign delimits the string from the numeric data in this file. Printing out `head` reveals that there are now two columns.

Data Manipulation and Analysis: Pandas

```
In [12]:   head
Out[12]:
                                         0        1
0       Wavelength [vacuum] (nm)      532,,
1       Refractive index of solvent   1.33,,
2       Refractive index of particles 1.59,,
3       Diameter of particles (microns) 0.5,,
```

This still isn't quite what we want, as the second column consists of numbers followed by two commas, which are unwanted remnants of the csv file. We get rid of the commas by declaring the comma to be a "comment" character (the symbol # is the default comment character in Python). We do this by introducing the keyword `comment`, as illustrated here:

```
In [13]:   head = pd.read_table('ScatMieData.csv', sep='=',
                                 nrows=4, comment=',',
                                 header=None)
```

Now typing `head` gives numbers without the trailing commas:

```
In [14]:   head
Out[14]:
                                         0        1
0       Wavelength [vacuum] (nm)      532.00
1       Refractive index of solvent    1.33
2       Refractive index of particles  1.59
3       Diameter of particles (microns) 0.50
```

Printing out individual elements of the two columns shows that the elements of column 0 are strings while the elements of column 1 are are floating point numbers, which is the desired result.

```
In [15]:   head[0][0]
Out[15]:   'Wavelength [vacuum] (nm) '

In [16]:   head[1][:]
Out[16]:
0     532.00
1       1.33
2       1.59
3       0.50
Name: 1, dtype: float64
```

Naming columns manually

If you prefer for the columns to be labeled by descriptive names instead of numbers, you can use the keyword `names` to provide names for the columns.

```
In [17]: head = pd.read_table('ScatMieData.csv', sep='=',
    ...:                       nrows=4, comment=',',
    ...:                       names=['property', 'value'])

Out[17]: head
                              property    value
0              Wavelength [vacuum] (nm)   532.00
1             Refractive index of solvent   1.33
2            Refractive index of particles  1.59
3            Diameter of particles (microns) 0.50

In [18]: head['property'][2]
Out[18]: 'Refractive index of particles '

In [19]: head['value'][2]
Out[19]: 1.5900000000000001
```

We can use what we have learned here to read data from the data file and then plot it, as shown in Fig. 10.2. Here is the code that produces the plot shown in Fig. 10.2.

Code: chapter10/programs/ScatMiePlot.py

```
1  import numpy as np
2  import matplotlib.pyplot as plt
3  import pandas as pd
4
5  # Read in data
6  head = pd.read_table('ScatMieData.csv', sep='=', nrows=4,
7                       comment=',', header=None)
8  scat = pd.read_csv('ScatMieData.csv', skiprows=4)
9
10 theta = (180./np.pi)*np.arccos(scat.Cos_theta)
11
12 plt.close('all')
13 fig, ax = plt.subplots(figsize=(6, 4))
14
15 ax.semilogy(theta, scat.F1, 'o', color='black', label="F1")
16 ax.semilogy(theta, scat.F2, 's', mec='black', mfc='white',
17             zorder=-1, label="F2")
18 ax.legend(loc="lower left")
19 ax.set_xlabel("theta (degrees)")
20 ax.set_ylabel("intensity")
21 for i in range(4):
22     ax.text(0.95, 0.9-i/18, "{} = {}"
23             .format(head[0][i], head[1][i]),
24             ha='right', fontsize=10, transform=ax.transAxes)
25 fig.tight_layout()
26 fig.savefig('ScatMiePlot.pdf')
```

Data Manipulation and Analysis: Pandas 247

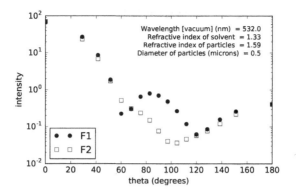

Figure 10.2 Plotting data from csv file read by Pandas routines.

10.1.2 Reading from text files

Perhaps the most common form of data file is simply a text file consisting of columns of data separated by spaces, tabs, or some other character. The workhorse for reading these types of files is `read_table()`, which we introduced, albeit briefly, in the previous section. In a nutshell, `read_table()` is exactly the same as `read_csv()` except it adds the keyword `sep`, which allows you to specify how columns of data are separated. In fact, `read_table(sep=',')` is completely equivalent to `read_csv()`.

Let's consider data about the planets stored in a text file with the data columns separated by spaces, as shown here.

Data: chapter10/programs/planetData.txt

```
planet      distance      mass   gravity  diameter      year
Mercury         0.39     0.055      0.38      0.38      0.24
Venus           0.72      0.82      0.91      0.95      0.62
Earth           1.00      1.00      1.00      1.00      1.00
Mars            1.52      0.11      0.38      0.53      1.88
Jupiter         5.20       318      2.36      11.2      11.9
Saturn          9.58        95      0.92      9.45        29
Uranus          19.2        15      0.89      4.01        84
Neptune         30.0        17      1.12      3.88       164
Pluto           39.5    0.0024     0.071      0.19       248
```

The quantities in the table are referenced to Earth. Each column is separated from the previous column by a variable number of spaces. The way Pandas handles this is by setting `sep='Series+'`.

```
In [20]: planets = pd.read_table('planetData.txt', sep='\s+')

In [21]: planets
Out[21]:
    planet  distance      mass  gravity  diameter    year
0  Mercury      0.39    0.0550    0.380      0.38    0.24
1    Venus      0.72    0.8200    0.910      0.95    0.62
2    Earth      1.00    1.0000    1.000      1.00    1.00
3     Mars      1.52    0.1100    0.380      0.53    1.88
4  Jupiter      5.20  318.0000    2.360     11.20   11.90
5   Saturn      9.58   95.0000    0.920      9.45   29.00
6   Uranus     19.20   15.0000    0.890      4.01   84.00
7  Neptune     30.00   17.0000    1.120      3.88  164.00
8    Pluto     39.50    0.0024    0.071      0.19  248.00
```

Of course, columns are frequently delimited in other ways. In §10.1.1, we wrote `sep='='` to use the equals sign as a column separator. For tab-delimited columns, use `sep='\t'`. For white-space-delimited columns (any combination of spaces and tabs), use `delim_whitespace=True` (and omit the `sep` keyword).

Notice that Pandas added a numerical index (the first column) to designate the rows in Out[21]. However, you might prefer to use the planet name rather than a number as the index. We can do this with the following code.

```
In [22]: planets = pd.read_table('planetData.txt', sep='\s+',
    ...:                         index_col='planet')

In [23]: planets
Out[23]:
         distance      mass  gravity  diameter    year
planet
Mercury      0.39    0.0550    0.380      0.38    0.24
Venus        0.72    0.8200    0.910      0.95    0.62
Earth        1.00    1.0000    1.000      1.00    1.00
Mars         1.52    0.1100    0.380      0.53    1.88
Jupiter      5.20  318.0000    2.360     11.20   11.90
Saturn       9.58   95.0000    0.920      9.45   29.00
Uranus      19.20   15.0000    0.890      4.01   84.00
Neptune     30.00   17.0000    1.120      3.88  164.00
Pluto       39.50    0.0024    0.071      0.19  248.00

In [24]: planets['distance']['Saturn']
Out[24]: 9.5800000000000001
```

As you can see from the output above, the numerical index has been replaced by a string index, which here is the name of the planet.

By the way, had our original text data file omitted the name `planet`

Data Manipulation and Analysis: Pandas

for the first column so that there were labels for only five of the six columns, then Pandas would simply have assumed that the first column was the index column. That is, suppose the original data file had looked like this (`planet` heading omitted):

Data: chapter10/programs/planetDataA.txt

```
          distance      mass   gravity  diameter     year
Mercury       0.39     0.055      0.38      0.38     0.24
Venus         0.72      0.82      0.91      0.95     0.62
Earth         1.00      1.00      1.00      1.00     1.00
Mars          1.52      0.11      0.38      0.53     1.88
Jupiter       5.20       318      2.36      11.2     11.9
Saturn        9.58        95      0.92      9.45       29
Uranus        19.2        15      0.89      4.01       84
Neptune       30.0        17      1.12      3.88      164
Pluto         39.5    0.0024     0.071      0.19      248
```

Now we read in this file without any designation of the index column.

```
In [25]: planets = pd.read_table('planetDataA.txt', sep='\s+')

In [26]: planets
Out[26]:
          distance       mass   gravity  diameter     year
Mercury       0.39     0.0550     0.380      0.38     0.24
Venus         0.72     0.8200     0.910      0.95     0.62
Earth         1.00     1.0000     1.000      1.00     1.00
Mars          1.52     0.1100     0.380      0.53     1.88
Jupiter       5.20   318.0000     2.360     11.20    11.90
Saturn        9.58    95.0000     0.920      9.45    29.00
Uranus       19.20    15.0000     0.890      4.01    84.00
Neptune      30.00    17.0000     1.120      3.88   164.00
Pluto        39.50     0.0024     0.071      0.19   248.00
In [27]: planets['mass']['Neptune']
Out[27]: 17.0
```

Table 10.1 summarizes some of the keyword arguments we used in the `read` functions we discussed above.

Keyword	Value	Description
sep	"\s+"	variable space-delimited data
	"\t"	tab-delimited data
	str	string-delimited data
delim_whitespace	True	mixed tab/space-delimited data
comment	*str*	set comment symbol (change from #)

Table 10.1 Basic keywords and values for use with Pandas `read` functions.

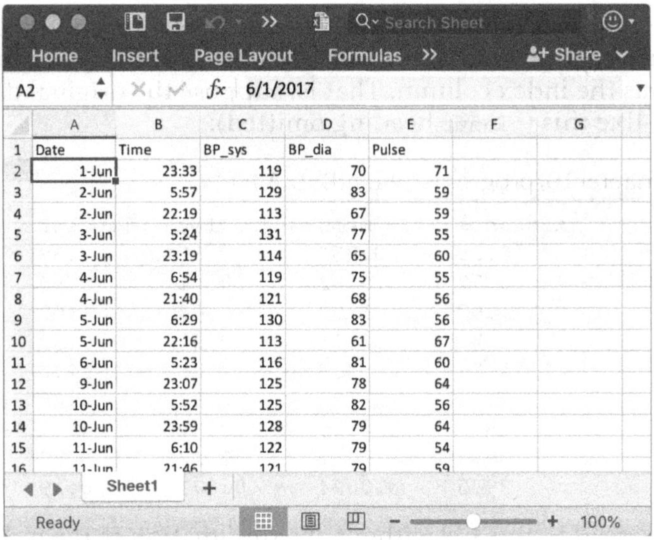

	A	B	C	D	E	F	G
1	Date	Time	BP_sys	BP_dia	Pulse		
2	1-Jun	23:33	119	70	71		
3	2-Jun	5:57	129	83	59		
4	2-Jun	22:19	113	67	59		
5	3-Jun	5:24	131	77	55		
6	3-Jun	23:19	114	65	60		
7	4-Jun	6:54	119	75	55		
8	4-Jun	21:40	121	68	56		
9	5-Jun	6:29	130	83	56		
10	5-Jun	22:16	113	61	67		
11	6-Jun	5:23	116	81	60		
12	9-Jun	23:07	125	78	64		
13	10-Jun	5:52	125	82	56		
14	10-Jun	23:59	128	79	64		
15	11-Jun	6:10	122	79	54		
16	11-Jun	21:46	121	79	59		

Figure 10.3 Excel file containing blood pressure data.

10.1.3 Reading from an Excel file

Pandas can also read directly from Excel files (*i.e.*, with .xls or .xlsx extensions). We consider an Excel file that contains blood pressure and pulse data taken twice per day, early in the morning and late in the evening, over a period of several weeks. The top of the Excel file is shown in Fig. 10.3 (there are many more rows, which are not shown). The file contains five columns: the date, time, systolic blood pressure, diastolic blood pressure, and pulse. The blood pressures are reported in mm-Hg and the pulse rate in heartbeats/minute. The name of the Excel file is `BloodPressure.xlsx`.

Reading in data from an Excel file using Pandas is simple:

```
In [28]: bp = pd.read_excel('BloodPressure.xlsx',
                            usecols='A:E')
```

The keyword argument `usecols='A:E'` tells Pandas to read in only columns A through E; data in any other columns are ignored. Had we wanted to read in only the pulse and not the blood pressure data, we could have written `usecols='A:B, E'` for the keyword argument. But as written, Pandas reads columns A through E into a DataFrame object named `bp`, whose structure we can see by typing `bp`:

```
In [29: bp
Out[29]:
```

```
         Date      Time  BP_sys  BP_dia  Pulse
0　 2017-06-01  23:33:00     119      70     71
1　 2017-06-02  05:57:00     129      83     59
2　 2017-06-02  22:19:00     113      67     59
3　 2017-06-03  05:24:00     131      77     55
4　 2017-06-03  23:19:00     114      65     60
5　 2017-06-04  06:54:00     119      75     55
6　 2017-06-04  21:40:00     121      68     56
7　 2017-06-05  06:29:00     130      83     56
8　 2017-06-05  22:16:00     113      61     67
9　 2017-06-06  05:23:00     116      81     60
10  2017-06-09  23:07:00     125      78     64
    .
    .
    .
```

Of course, we could have used the `read_excel` function to read in the Excel file shown in Fig. 10.1 instead of the `read_csv` function we employed in §10.1.1. Using `read_excel`, we can skip the first four rows using the `skiprows` keyword, just as we did using the `read_csv` function.

Function	Description
read_table	workhorse: read tabular data from a text file
read_csv	read tabular data from a comma separated file
read_excel	read tabular data from an Excel file
read_clipboard	read data copied from web page to clipboard
read_fwf	read data in fixed-width columns w/o delimiters

Table 10.2 Summary of Pandas functions to read tabular data files.

10.2 Dates and Times in Pandas

Pandas has special tools for handling dates and times. These tools make use of the Python library datetime, which defines, among other things, a useful *datetime* object. A datetime object stores, as its name implies, a precise moment in time. To see how this works, let's import the datetime library and then get the current value of `datetime`:

```
In [1]: import datetime as dt

In [2]: t0 = dt.datetime.now()
```

```
In [3]: t0
Out[3]: datetime.datetime(2017, 7, 21, 8, 17, 24, 241916)
```

The datetime object returns the year, month, day, hour, minute, second, and microsecond. You can format a datetime object for printing using the strftime method of the datetime library in various ways:

```
In [4]: t0.strftime('%Y-%m-%d')
Out[4]: '2017-07-21'

In [5]: t0.strftime('%d-%m-%Y')
Out[5]: '21-07-2017'

In [6]: t0.strftime('%d-%b-%Y')
Out[6]: '21-Jul-2017'

In [7]: t0.strftime('%H:%M:%S')
Out[7]: '09:00:15'

In [8]: t0.strftime('%Y-%B-%d %H:%M:%S')
Out[8]: '2017-July-21 09:00:15'
```

You can construct almost any format you want.

In the previous section, we read blood pressure and pulse information into a DataFrame with the date and time as separate variables (*i.e.*, columns). It is usually more convenient within Pandas to have these combined as a single datetime object. While it is possible to combine the bp['Date'] and bp['Time'] columns in the bp DataFrame into a single datetime object after reading the original Excel file, it is much easier to read the two Excel columns *directly* into a datetime object using the keyword argument parse_dates, as illustrated here:

```
In [9]: bp = pd.read_excel('BloodPressure.xlsx',
   ...:                    usecols='A:E',
   ...:                    parse_dates=[['Date', 'Time']])

In [10]: bp
Out[10]:
            Date_Time  BP_sys  BP_dia  Pulse
0  2017-06-01 23:33:00     119      70     71
1  2017-06-02 05:57:00     129      83     59
2  2017-06-02 22:19:00     113      67     59
3  2017-06-03 05:24:00     131      77     55
4  2017-06-03 23:19:00     114      65     60
5  2017-06-04 06:54:00     119      75     55
6  2017-06-04 21:40:00     121      68     56
7  2017-06-05 06:29:00     130      83     56
8  2017-06-05 22:16:00     113      61     67
```

```
9   2017-06-06 05:23:00    116     81    60
10  2017-06-09 23:07:00    125     78    64
                .
                .
                .
```

The `parse_dates` keyword argument can also be used with the `read_csv` and `read_table` methods.

10.3 Data Structures: Series and DataFrame

Pandas has two principal data structures: *Series* and *DataFrame*. They form the basis for most activities using Pandas. We have already met the DataFrame in §10.1, and have learned a bit about its most basic properties and uses.

A Series object is a one-dimensional DataFrame.

Both Series and DataFrames use NumPy arrays extensively, but allow more versatile ways of indexing, as we have already seen for DataFrames. These more versatile ways of indexing facilitate convenient ways of combining and manipulating different data sets, and thus are at the heart of Pandas functionality. This added functionality can come at a performance price, a topic we briefly address below.

10.3.1 Series

A Pandas Series is a one-dimensional array-like data structure made up of a NumPy array and an associated array of data labels called the *index*. We can create a Series using the Pandas `Series` function, which turns a list, dictionary, or NumPy array into a Pandas Series. Here we use it to turn a list into a Series:

```
In [1]: ht = pd.Series([160.0-4.9*t*t for t in range(6)])

In [2]: ht
Out[2]:
0    160.0
1    155.1
2    140.4
3    115.9
4     81.6
5     37.5
dtype: float64
```

The IPython output displayed is in two columns, the index column on the left and the values of the Series on the right. The argument of the `Series` function can be a list, an iterator, or a NumPy array. In this case, the values of the Series are floating point numbers. The index goes from 0 to $N-1$, where N is the number of data points. Individual elements and slices are accessed in the same way as for lists and NumPy arrays.

```
In [3]: ht[2]
Out[3]: 140.40000000000001

In [4]: ht[1:4]
Out[4]:
1    155.1
2    140.4
3    115.9
dtype: float64
```

The entire array of values and indices can be accessed using the `values` and `index` attributes.

```
In [5]: ht.values
Out[5]: array([ 160. ,  155.1,  140.4,  115.9,   81.6,   37.5])

In [6]: ht.index
Out[6]: RangeIndex(start=0, stop=6, step=1)
```

Here, as can be seen from the outputs, the `ht.values` is a NumPy array and `ht.index` is an iterator.

Indexing Series works exactly as it does for DataFrames. So, unless otherwise specified, the index default is to go from 0 to $N-1$ where N is the size of the Series. However, other indexing schemes can be specified:

```
In [7]: heights = pd.Series([188, 157, 173, 169, 155],
   ...:                    index=['Jake', 'Sarah', 'Heather',
   ...:                           'Chris', 'Alex'])

In [8]: heights
Out[8]:
Jake       188
Sarah      157
Heather    173
Chris      169
Alex       155
dtype: int64
```

The Series `heights` bears a striking resemblance to a Python dictionary. Indeed, a Pandas Series can be converted to a dictionary using the `to_dict` method:

Data Manipulation and Analysis: Pandas

```
In [9]: htd = heights.to_dict()

In [10]: htd
Out[10]: {'Alex': 155, 'Chris': 169, 'Heather': 173,
          'Jake': 188, 'Sarah': 157}
```

A Python dictionary can be converted to a Pandas Series using the `Series` function:

```
In [11]: pd.Series(htd)
Out[11]:
Alex       155
Chris      169
Heather    173
Jake       188
Sarah      157
dtype: int64
```

Time series

One of the most common uses of Pandas series involves a *time* series in which the series is indexed by timestamps. For example, we can change the index of the series `ht` to consecutive days. First we create a time sequence:

```
In [12]: dtr = pd.date_range('2017-07-22', periods=6)
```

Then we set the index of `ht` to the time sequence:

```
In [13]: ht.index = dtr

In [14]: ht
Out[14]:
2017-07-22    160.0
2017-07-23    155.1
2017-07-24    140.4
2017-07-25    115.9
2017-07-26     81.6
2017-07-27     37.5
Freq: D, dtype: float64

In [15]: ht['2017-07-25']
Out[15]: 115.90000000000001

In [16]: ht['2017-07-23':'2017-07-26']
Out[16]:
2017-07-23    155.1
2017-07-24    140.4
2017-07-25    115.9
```

```
2017-07-26      81.6
Freq: D, dtype: float64
```

Note that you can slice time series indexed by dates, and that the slice range includes both the starting and ending dates.

Alternatively, the time series can be created in one step using the `Series` function of Pandas:

```
In [17]: htAlt = pd.Series([160.0-4.9*t*t for t in range(6)],
    ...:                    index=dtr)

In [18]: htAlt
Out[18]:
2017-07-22      160.0
2017-07-23      155.1
2017-07-24      140.4
2017-07-25      115.9
2017-07-26       81.6
2017-07-27       37.5
Freq: D, dtype: float64
```

10.3.2 DataFrame

We introduced the DataFrame, a two-dimensional spreadsheet-like data structure, in §10.1. There we learned that we can access a column of a DataFrame using column labels, like `planets['mass']` to get the masses of all the planets in the `mass` column or a single DataFrame cell, like `planets['mass']['Mars']`, to get the mass of a single planet. However, this method of retrieving 2-dimensional data in a DataFrame is not very efficient and can be excruciatingly slow if one needs to sift through a large quantity of data. The recommended scheme is to use the `iloc` and `loc` methods, which are faster and more versatile.

Let's explore the indexing methods of Pandas by reloading the blood pressure Excel file:

```
In [19]: bp = pd.read_excel('BloodPressure.xlsx',
    ...:                    usecols='A:E',
    ...:                    parse_dates=[['Date', 'Time']])

In [20]: bp.head(4)   # print out first 5 lines of bp
Out[18]:
             Date_Time  BP_sys  BP_dia  Pulse
0  2017-06-01 23:33:00     119      70     71
1  2017-06-02 05:57:00     129      83     59
2  2017-06-02 22:19:00     113      67     59
```

Data Manipulation and Analysis: Pandas   257

```
3 2017-06-03 05:24:00        131      77       55

In [21]: bp.tail(4)    # print out last 5 lines of bp
Out[21]:
              Date_Time   BP_sys  BP_dia  Pulse
44  2017-07-15 22:57:00      109      63     62
45  2017-07-16 06:45:00      124      78     47
46  2017-07-16 22:15:00      121      74     58
47  2017-07-17 06:22:00      113      79     57
```

The `iloc` method

The `iloc` method indexes the DataFrame by row and column *number* (note the order: [row, column]—it's opposite to what we used before):

```
In [22]: bp.iloc[0, 2]
Out[22]: 70
```

The row and column numbers use the usual Python zero-based indexing scheme. The usual slicing syntax applies:

```
In [23]: bp.iloc[45, 0:3]
Out[23]:
Date_Time      2017-07-16 06:45:00
BP_sys                         124
BP_dia                          78
Name: 45, dtype: object
```

Suppose we set the `Date_Time` column to be the index:

```
In [23]: bp = bp.set_index('Date_Time')

In [24]: bp.head(4)
Out[24]:
                      BP_sys  BP_dia  Pulse
Date_Time
2017-06-01 23:33:00      119      70     71
2017-06-02 05:57:00      129      83     59
2017-06-02 22:19:00      113      67     59
2017-06-03 05:24:00      131      77     55

In [25]: bp.iloc[1, 0:2]
Out[25]:
BP_sys     129
BP_dia      83
Name: 2017-06-02 05:57:00, dtype: int64
```

Notice that column 0 is now the `BP_sys` column; the index column is not counted. As with lists and NumPy arrays, an index of −1 signifies the last element, −2 the next to last element, and so on.

The `loc` method

The `loc` method is an extremely versatile tool for indexing DataFrames. At first look, it seems simply like an alternative syntax for accomplishing the same thing you might have done with the `iloc` method:

```
In [26]: bp.loc['2017-06-02 05:57:00', 'BP_sys':'BP_dia']
Out[26]:
BP_sys    129
BP_dia     83
Name: 2017-06-02 05:57:00, dtype: int64
```

But it can do much more. Suppose, for example, we wanted to know if there were significant differences in the blood pressure readings taken in the morning from those taken in the evening. The `loc` method has, as a part of its functionality, the ability to select data based on conditions. We illustrate this by separating our pulse data into those measurements taken in the morning and those taken in the evening.

```
In [27]: PulseAM = bp.loc[bp.index.hour<12, 'Pulse']

In [28]: PulsePM = bp.loc[bp.index.hour>=12, 'Pulse']
```

Here we set the row entry of the `loc` method to a *condition* on the value of the time index, whether it is before or after noon (less than 12 hours or not). This is sometimes referred to as *conditional* or *Boolean indexing*.

You can specify quite complicated conditions if you wish. For example, suppose we wanted a list of all the planets that were more massive than the Earth but nevertheless had a smaller gravitational force at their surface. Using our DataFrame `planets` introduced previously, we could list those planets as follows:

```
In [29]: planets = pd.read_table('planetData.txt', sep='\s+',
    ...:                         index_col='planet')

In [30]: planets.loc[(planets['mass'] > 1.0) &
    ...:             (planets['gravity'] < 1.0)]
Out[30]:
        distance  mass  gravity  diameter  year
planet
Saturn      9.58  95.0     0.92      9.45  29.0
Uranus     19.20  15.0     0.89      4.01  84.0
```

The parentheses within the `loc` method are needed to define the order in which the logical operations are applied.

If we don't want to see all of the columns, we can specify the columns as well as the rows:

```
In [31]: planets.loc[(planets.mass > 1.0) &
   ...:                (planets.gravity < 1.0),
   ...:                'mass':'gravity']

Out[31]:
         mass   gravity
planet
Saturn   95.0    0.92
Uranus   15.0    0.89
```

Note that here we could have used `planets.mass` instead of `planets['mass']` and `planets.gravity` instead of `planets['gravity']`. Either works.

Creating a DataFrame

Up until now, the DataFrames we worked with were created for us when we read data in from a text, csv, or Excel file. Alternatively, you can create a DataFrame using the Pandas `DataFrame` routine. As input you can use nearly any list-like object, including a list, a NumPy array, or a dictionary. Perhaps the simplest way is using a dictionary.

```
In [32]: optmat = {'mat': ['silica', 'titania', 'PMMA', 'PS'],
   ...:            'index': [1.46, 2.40, 1.49, 1.59],
   ...:            'density': [2.03, 4.2, 1.19, 1.05]}

In [34]: omdf = pd.DataFrame(optmat)
Out[34]:
   density  index      mat
0     2.03   1.46   silica
1     4.20   2.40  titania
2     1.19   1.49     PMMA
3     1.05   1.59       PS
```

You can coerce the columns to appear in some desired order using the `columns` keyword argument.

```
In [35]: omdf = pd.DataFrame(optmat, columns=['mat', 'density',
                                              'index'])

In [36]: omdf
Out[36]:
       mat  density  index
0   silica     2.03   1.46
1  titania     4.20   2.40
2     PMMA     1.19   1.49
3       PS     1.05   1.59
```

We can also create a DataFrame with empty columns and fill in the data later.

```
In [37]: omdf1 = pd.DataFrame(index=['silica', 'titania',
                                    'PMMA', 'PS'],
                              columns={'density', 'index'})
```

```
In [38]: omdf1
Out[38]:
         index density
silica     NaN     NaN
titania    NaN     NaN
PMMA       NaN     NaN
PS         NaN     NaN
```

The index and column names are indicated but there is no data. The empty data columns are indicated by NaN (not-a-number). We can fill in the empty entries as follows:

```
In [39]: omdf1.loc['PS', ('index', 'density')] = (1.05, 1.59)
```

```
In [40]: omdf1
Out[40]:
         index density
silica     NaN     NaN
titania    NaN     NaN
PMMA       NaN     NaN
PS        1.05    1.59
```

Let's check the data types in our DataFrame.

```
In [41]: omdf1.dtypes
Out[41]:
index      object
density    object
dtype: object
```

The data types for the index and density columns were set to be object when the DataFrame was created because we gave these columns no data. Now that we have entered the data, we would prefer that the index and density columns be the float data type. To do so, we explicitly set the data type.

```
In [42]: omdf1[['index', 'density']] = omdf1[
                 ['index', 'density']].apply(pd.to_numeric)
```

```
In [43]: omdf1.dtypes
Out[43]:
index      float64
density    float64
dtype: object
```

Now the columns are properly typed as floating point numbers while the overall DataFrame is an object.

10.4 Getting Data from the Web

Pandas has extensive tools for scraping data from the web. Here we illustrate one of the simpler cases, reading a csv file from a web site. The Bank of Canada publishes the daily exchange rates between the Canadian dollar and a couple dozen international currencies. We would like to download these data and print out the results as a simple table. Here we employ Pandas's usual `read_csv` function using its `url` keyword argument to specify the web address of the csv file we want to read. To follow the code, download the csv file manually using the url defined in line 6 of the program below and then open it using a spreadsheet program like Excel.

To obtain all the data we want, we read the csv file twice. In line 8, we call `read_csv` to read into a DataFrame `rates` the exchange rates for the different currencies over a range of dates that extends from any start date (after 2017-01-03, the earliest date for which the site supplies data) up to the most recent business day. The rates are indexed by date (*e.g.*, `'2018-04-23'`) with each column corresponding to a different currency.

The header for the exchange rates, which consists of codes for each exchange rate, begins on line 40 so we skip the first 39 rows. In line 11 of the program, we get the number of days and the number of currencies downloaded from the shape of the DataFrame.

We read the file again on lines 13–14 to get keys for the codes for the various currencies used in the DataFrame. We use the number of currencies determined in line 8 to determine the number of lines to read. Lines 15–16 strip off some extraneous verbiage in the keys.

Code: chapter10/programs/urlRead.py

```
import pandas as pd

url1 = 'http://www.bankofcanada.ca/'
url2 = 'valet/observations/group/FX_RATES_DAILY/csv?start_date='
start_date = '2017-01-03'     # Earliest start date is 2017-01-03
url = url1+url2+start_date    # Complete url to download csv file
# Read in rates for different currencies for a range of dates
rates = pd.read_csv(url, skiprows=39, index_col='date')
rates.index = pd.to_datetime(rates.index)   # assures data type
```

```
10  # Get number of days & number of currences from shape of rates
11  days, currencies = rates.shape
12  # Read in the currency codes & strip off extraneous part
13  codes = pd.read_csv(url, skiprows=10, usecols=[0, 2],
14                     nrows=currencies)
15  for i in range(currencies):
16      codes.iloc[i, 1] = codes.iloc[i, 1].split(' to Canadian')[0]
17  # Report exchange rates for the most most recent date available
18  date = rates.index[-1]    # most recent date available
19  print('\nCurrency values on {0}'.format(date))
20  for (code, rate) in zip(codes.iloc[:, 1], rates.loc[date]):
21      print("{0:20s}  Can$ {1:8.6g}".format(code, rate))
```

Using the `index` attribute for Pandas DataFrames, line 18 sets the date for which the currency exchange data will be displayed, in this case, the most recent date in the file. Running the program produces the desired output:

```
In [1]: run urlRead.py

Currency values on 2018-09-18
Australian dollar      Can$   0.9367
Brazilian real         Can$   0.3143
Chinese renminbi       Can$   0.1893
European euro          Can$   1.5179
Hong Kong dollar       Can$   0.1656
Indian rupee           Can$   0.01782
Indonesian rupiah      Can$   8.7e-05
Japanese yen           Can$   0.01157
Malaysian ringgit      Can$   0.3136
Mexican peso           Can$   0.06918
New Zealand dollar     Can$   0.8555
Norwegian krone        Can$   0.1593
Peruvian new sol       Can$   0.3929
Russian ruble          Can$   0.01928
Saudi riyal            Can$   0.3464
Singapore dollar       Can$   0.9477
South African rand     Can$   0.08743
South Korean won       Can$ 0.001156
Swedish krona          Can$   0.1459
Swiss franc            Can$   1.349
Taiwanese dollar       Can$   0.04217
Thai baht              Can$   0.03989
Turkish lira           Can$   0.204
UK pound sterling      Can$   1.708
US dollar              Can$   1.2992
Vietnamese dong        Can$   5.6e-05
```

What we have done here illustrates only one simple feature of Pandas for scraping data from the web. Many more web-scraping tools exist within Pandas. They are extensive and powerful, and can be used in concert with other packages, such as `urllib3`, to extract almost any data that exists on the web.

10.5 Extracting Information from a DataFrame

Once we have our data organized in a DataFrame, we can employ the tools of Pandas to extract and summarize the data it contains in a variety of ways. Here we will illustrate a few of Pandas' tools using the `planets` and `bp` DataFrames we introduced in §10.1.2 and §10.1.3. We read them in again for good measure:

```
In [1]: planets = pd.read_table('planetData.txt', sep='\s+',
   ...:                          index_col='planet')

In [2]: bp = pd.read_excel('BloodPressure.xlsx',
   ...:                     usecols='A:E',
   ...:                     parse_dates=[['Date', 'Time']])

In [3]: bp = bp.set_index('Date_Time')

In [4]: planets
Out[4]:
         distance      mass   gravity   diameter      year
planet
Mercury      0.39    0.0550     0.380       0.38      0.24
Venus        0.72    0.8200     0.910       0.95      0.62
Earth        1.00    1.0000     1.000       1.00      1.00
Mars         1.52    0.1100     0.380       0.53      1.88
Jupiter      5.20  318.0000     2.360      11.20     11.90
Saturn       9.58   95.0000     0.920       9.45     29.00
Uranus      19.20   15.0000     0.890       4.01     84.00
Neptune     30.00   17.0000     1.120       3.88    164.00
Pluto       39.50    0.0024     0.071       0.19    248.00
```

Note that we have set the `planet` column to be the index variable in the `planets` DataFrame.

Pandas can readily sort data. For example, to list the planets in order of increasing mass, we write:

```
In [5]: planets.sort_values(by='mass')
Out[5]:
         distance      mass   gravity   diameter      year
planet
Pluto       39.50    0.0024     0.071       0.19    248.00
```

Mercury	0.39	0.0550	0.380	0.38	0.24
Mars	1.52	0.1100	0.380	0.53	1.88
Venus	0.72	0.8200	0.910	0.95	0.62
Earth	1.00	1.0000	1.000	1.00	1.00
Uranus	19.20	15.0000	0.890	4.01	84.00
Neptune	30.00	17.0000	1.120	3.88	164.00
Saturn	9.58	95.0000	0.920	9.45	29.00
Jupiter	5.20	318.0000	2.360	11.20	11.90

To produce the same table but from highest to lowest mass, use the keyword argument `ascending=False`.

We can use conditional indexing to get a list of all the planets with gravitational acceleration larger than Earth's.

```
In [6]: planets[planets['gravity']>1]
Out[6]:
         distance   mass   gravity  diameter   year
planet
Jupiter      5.2   318.0      2.36     11.20   11.9
Neptune     30.0    17.0      1.12      3.88  164.0
```

It's worth parsing In [6] to better understand how it works. Suppose we had typed just what is inside the outermost brackets:

```
In [7]: planets['gravity']>1
Out[7]:
planet
Mercury    False
Venus      False
Earth      False
Mars       False
Jupiter     True
Saturn     False
Uranus     False
Neptune     True
Pluto      False
Name: gravity, dtype: bool
```

We get the logical (Boolean) truth values for each entry. Thus, writing `planets[planets['gravity']>1]` lists the DataFrame only for those entries where the Boolean value is `True`.

Suppose we would like to find the volume V of each of the planets and add the result to our `planets` DataFrame. Using the formula $V = \frac{1}{6}\pi d^2$, where d is the diameter, we simply write

```
In [8]: planets['volume'] = pi * planets['diameter']**3 / 6.0

In [9]: planets
```

Data Manipulation and Analysis: Pandas 265

```
Out[9]:
         distance      mass  gravity  diameter    year    volume
planet
Mercury      0.39    0.0550    0.380      0.38    0.24    0.0287
Venus        0.72    0.8200    0.910      0.95    0.62    0.4489
Earth        1.00    1.0000    1.000      1.00    1.00    0.5236
Mars         1.52    0.1100    0.380      0.53    1.88    0.0780
Jupiter      5.20  318.0000    2.360     11.20   11.90  735.6186
Saturn       9.58   95.0000    0.920      9.45   29.00  441.8695
Uranus      19.20   15.0000    0.890      4.01   84.00   33.7623
Neptune     30.00   17.0000    1.120      3.88  164.00   30.5840
Pluto       39.50    0.0024    0.071      0.19  248.00    0.0036
```

Let's look at the blood pressure DataFrame, where we have set the Date_Time column to be the index variable in the bp DataFrame.

```
In [10]: bp
                     BP_sys  BP_dia  Pulse
Date_Time
2017-06-01 23:33:00     119      70     71
2017-06-02 05:57:00     129      83     59
2017-06-02 22:19:00     113      67     59
2017-06-03 05:24:00     131      77     55
2017-06-03 23:19:00     114      65     60
         .
         .
         .
```

Note that we have set the planet and Date_Time columns to be the index variables in the planets and bp DataFrames.

Pandas can calculate standard statistical quantities for the data in a DataFrame.

```
In [8]: bp['BP_sys'].mean()       # average systolic pressure
Out[8]: 119.27083333333333

In [9]: bp['BP_sys'].max()        # maximum systolic pressure
Out[9]: 131

In [10]: bp['BP_sys'].min()       # minimum systolic pressure
Out[10]: 105

In [11]: bp['BP_sys'].count()     # num (non-null) of entries
Out[11]: 48
```

The statistical methods can even act on dates, if doing so makes sense.

```
In [12]:   bp.index.min()         # starting datetime
Out[12]: Timestamp('2017-06-01 23:33:00')
```

```
In [13]: bp.index.max()              # ending datetime
Out[13]: Timestamp('2017-07-17 06:22:00')
```

Note that here we used `bp.index` and not `bp.['Date_Time']`, as we previously set `'Date_Time'` to be the index of `bp`. Time differences can also be calculated quite simply:

```
In [14]: bp.index.max()-bp.index.min()
Out[14]: Timedelta('45 days 06:49:00')
```

Function	Description	Function	Description
min	minimum	cummin	cumulative minimum
max	maximum	cummax	cumulative maximum
mean	mean	skew	skewness
median	median	kurt	kurtosis
mode	mode	quantile	quantile
var	variance	mad	mean abs deviation
std	standard deviation	sem	standard error of mean
abs	absolute value	count	num non-null entries
sum	sum	cumsum	cumulative sum
prod	product	cumprod	cumulative product
describe	count, mean, std, min, max, & percentiles		

Table 10.3 Statistical methods for Pandas DataFrame and Series.

We can combine these methods with the conditional indexing of the last section to answer some interesting questions. For example, are there systematic differences in the blood pressure and pulse readings in the morning and the evening? Let's use what we've learned to find out. Previously we had:

```
In [15]: PulseAM = bp.loc[bp.index.hour<12, 'Pulse']

In [16]: PulsePM = bp.loc[bp.index.hour>=12, 'Pulse']
```

Now let's look at some averages and fluctuations about the mean:

```
In [17]: precision 3
Out[17]:: '%.3f'

In [18]: PulseAM.mean(), PulseAM.std(), PulseAM.sem()
Out[18]:: (57.586, 5.791, 1.075)

In [19]: PulsePM.mean(), PulsePM.std(), PulsePM.sem()
Out[19]:: (61.789, 4.939, 1.133)
```

Data Manipulation and Analysis: Pandas   267

We see that the average morning pulse of 57.6 is lower than the average evening pulse of 61.8. The difference of 4.2 is greater than the standard error of the mean of about 1.1, which means the difference is significant, even though the morning and evening pulse distributions overlap each other to a significant extent, as indicated by the standard deviations of around 5.

Finally, we can get histogram data on frequency.

10.6 Plotting with Pandas

In Chapter 6, we introduced the matplotlib plotting package, which provides an extensive framework for plotting within Python. Pandas builds on the matplotlib package, adding some functionality peculiar to Pandas.

One notable change is that when plotting data from a Pandas Series or DataFrame, matplotlib's `plot` function will use the index as the *x* data if the *x* data is not otherwise specified. For example, we can get a graphical display, shown in Fig. 10.4, of the relative gravity of each planet from the `planets` DataFrame with the following simple commands:

```
In [1]: planets['gravity'].plot.bar(color='C0')
Out[1]: <matplotlib.axes._subplots.AxesSubplot at 0x11de29400>

In [2]: ylabel('relative gravity')
Out[2]: Text(42.5972,0.5,'relative gravity')

In [3]: tight_layout()
```

Pandas allows us to write plotting commands in a new way, where a matplotlib plotting function is now a DataFrame method. Here we use `plot` as a method of `planets['gravity']`. We further specify a `bar` (histogram) plot with `bar` as a method `plot`. The *y*-axis is specified by choosing the desired column(s) of the DataFrame, in this case `gravity`, and the *x*-axis is taken to be the DataFrame index unless otherwise specified. Notice how each bar is neatly labeled with its corresponding planet index. We could have made a plot with horizontal instead of vertical bars using the `barh` method in place of `bar`. Try it out!

Let's look at another example, this time using our `bp` DataFrame. First, let's plot it using the conventional matplotlib syntax,

```
In [1]: plot(bp)
Out[1]::
```

```
[<matplotlib.lines.Line2D at 0x13667b278>,
 <matplotlib.lines.Line2D at 0x1366aecf8>,
 <matplotlib.lines.Line2D at 0x1366b7080>]
```

which produces the graph on the left in Fig. 10.5. The three traces correspond to the systolic pressure, the diastolic pressure, and the pulse, and are plotted as a function of the time (date), which is the index of the `bp` DataFrame. Since the x-array is not specified, the index variable, the date, is used. However, the dates are not so nicely formatted and run into each other.

Alternatively, we can graph this using `plot` as a DataFrame method:

```
In [2]: bp.plot()
Out[2]: <matplotlib.axes._subplots.AxesSubplot at 0x1527ddd240>
```

The result, shown on the right in Fig. 10.5, is a more nicely formatted plot, where the dates labeling the x-axis are automatically tilted so that they don't run into each other, and a legend is produced, which identifies the different traces.

Figure 10.6 shows these same data in a more compelling and refined graph, putting together much of the analysis we have already developed using Pandas. Measurements made early in the morning and late in the evening are distinguished from each other using open and closed symbols. The morning and evening averages are indicated by horizontal lines that are annotated with the numerical averages and indicated using arrows. A more compete legend is supplied.

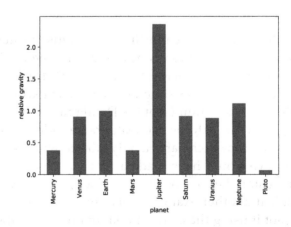

Figure 10.4 Relative gravity of different planets.

Data Manipulation and Analysis: Pandas

The code shows how Pandas and conventional matplotlib syntax can be used in concert with each other. The blood pressure and pulse data are plotted on separate graphs sharing a common time axis. The code that produces Fig. 10.6 is listed below. Note that which plot is chosen, ax1 or ax2, is indicated using the keyword argument ax within the plot method belonging to the various data sets, sysPM, ..., PulsePM. Finally, matplotlib's dates package is used to format the *x*-axis.

Much more information is available at the Pandas web site, which gives details about all of Pandas' plotting commands.

Code: chapter10/programs/BloodPressure.py

```
 1  import matplotlib.pyplot as plt
 2  import pandas as pd
 3  import matplotlib.dates as mdates
 4  from datetime import datetime
 5
 6  # Read in data
 7  bp = pd.read_excel('BloodPressure.xlsx', usecols='A:E',
 8                     parse_dates=[['Date', 'Time']])
 9  bp = bp.set_index('Date_Time')
10  # Divide data into AM and PM sets
11  diaAM = bp.loc[bp.index.hour < 12, 'BP_dia']
12  diaPM = bp.loc[bp.index.hour >= 12, 'BP_dia']
13  sysAM = bp.loc[bp.index.hour < 12, 'BP_sys']
14  sysPM = bp.loc[bp.index.hour >= 12, 'BP_sys']
15  PulseAM = bp.loc[bp.index.hour < 12, 'Pulse']
16  PulsePM = bp.loc[bp.index.hour >= 12, 'Pulse']
17  # Set up figure with 2 subplots and plot BP data
18
```

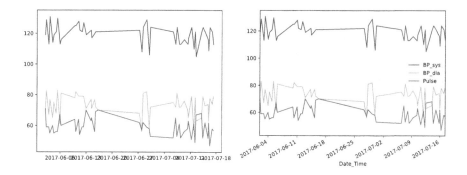

Figure 10.5 Crude plots of the bp DataFrame.

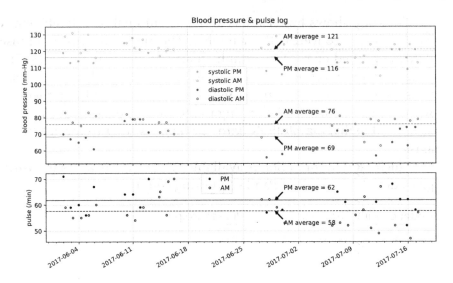

Figure 10.6 Blood pressure data from an Excel file.

```
19  fig, (ax1, ax2) = plt.subplots(2, 1, sharex=True,
20                      gridspec_kw={'height_ratios': [2, 1]},
21                      figsize=(10, 6))
22  fig.subplots_adjust(left=0.065, right=0.99, hspace=0.06)
23  sysPM.plot(ax=ax1, marker='o', ms=3, lw=0, color='C1',
24          label='systolic PM')
25  sysAM.plot(ax=ax1, marker='o', ms=3, lw=0, color='C1',
26          mfc='white', label='systolic AM')
27  diaPM.plot(ax=ax1, marker='o', ms=3, lw=0, color='C0',
28          label='diastolic PM')
29  diaAM.plot(ax=ax1, marker='o', ms=3, lw=0, color='C0',
30          mfc='white', label='diastolic AM')
31  # Average values of blood pressures with arrows labeling them
32  dtlab = datetime(2017, 6, 29)
33  bpavgs = (sysAM.mean(), sysPM.mean(), diaAM.mean(),
34          diaPM.mean())
35  ytext = ('bottom', 'top')
36  tavgs = ('AM average = {0:0.0f}'.format(bpavgs[0]),
37          'PM average = {0:0.0f}'.format(bpavgs[1]),
38          'AM average = {0:0.0f}'.format(bpavgs[2]),
39          'PM average = {0:0.0f}'.format(bpavgs[3]))
40  aprops = dict(facecolor='black', width=1, headlength=5,
41          headwidth=5)
42  for i, bpa in enumerate(bpavgs):
43      ax1.annotate(tavgs[i], xy=(dtlab, bpa),
44              xytext=((15, (-1)**(i % 2)*15)),
45              textcoords='offset points',
```

Data Manipulation and Analysis: Pandas

```
46                     arrowprops=aprops, ha='left',
47                     va=ytext[i % 2])
48  # Lines indicating average blood pressures
49  ax1.axhline(y=sysPM.mean(), color='C1', lw=0.75, zorder=-1)
50  ax1.axhline(y=sysAM.mean(), color='C1', dashes=(5, 2),
51              lw=0.75, zorder=-1)
52  ax1.axhline(y=diaPM.mean(), color='C0', lw=0.75, zorder=-1)
53  ax1.axhline(y=diaAM.mean(), color='C0', dashes=(5, 2),
54              lw=0.75, zorder=-1)
55  # Formatting top graph
56  ax1.set_title('Blood pressure & pulse log')
57  ax1.set_ylabel('blood pressure (mm-Hg)')
58  ax1.legend(loc=(0.37, 0.43))
59  ax1.grid(dashes=(1, 2))
60  # Plot pulse
61  PulsePM.plot(ax=ax2, marker='o', ms=3, lw=0, color='k',
62               label='PM')
63  PulseAM.plot(ax=ax2, marker='o', ms=3, lw=0, color='k',
64               mfc='white', label='AM')
65  # Average values of pulse with arrows labeling them
66  Pulseavgs = (PulseAM.mean(), PulsePM.mean())
67  tavgs = ('AM average = {0:0.0f}'.format(Pulseavgs[0]),
68           'PM average = {0:0.0f}'.format(Pulseavgs[1]))
69  for i, pulse in enumerate(Pulseavgs):
70      ax2.annotate(tavgs[i], xy=(dtlab, pulse),
71                   xytext=((15, -(-1)**(i)*15)),
72                   textcoords='offset points',
73                   arrowprops=aprops, ha='left',
74                   va=ytext[-i-1])
75
76  ax2.axhline(y=PulsePM.mean(), color='k', lw=0.75, zorder=-1)
77  ax2.axhline(y=PulseAM.mean(), color='k', dashes=(5, 2),
78              lw=0.75, zorder=-1)
79  # Formatting bottom graph
80  week = mdates.WeekdayLocator(byweekday=mdates.SU)
81  day = mdates.DayLocator()
82  ax2.xaxis.set_major_locator(week)
83  ax2.xaxis.set_minor_locator(day)
84  ax2.set_xlabel('')
85  ax2.set_ylabel('pulse (/min)')
86  ax2.legend(loc=(0.4, 0.7))
87  ax2.grid(dashes=(1, 2))
88
89  fig.tight_layout()
90  fig.show()
91  fig.savefig('./figures/BloodPressure.pdf')
```

10.7 Grouping and Aggregation

Pandas allows you to group data and analyze the subgroups in useful and powerful ways. The best way to understand what you can do is to work with an example. Here we will work with a data set that lists all the departures from Newark Liberty International Airport (EWR) on a particular (stormy) day. The data is stored in a csv file named `ewrFlights20180516.csv`, which we read into a DataFrame that we call ewr.

```
In [1]: ewr = pd.read_csv('ewrFlights20180516.csv')

In [2]: ewr.head()
Out[2]:
      Destination                 Airline   Flight Departure  \
0   Baltimore (BWI)    Southwest Airlines   WN 8512  12:09 AM
1   Baltimore (BWI)    Mountain Air Cargo   C2 7304  12:10 AM
2       Paris (ORY)  Norwegian Air Shuttle   DY 7192  12:30 AM
3       Paris (ORY)       euroAtlantic Airways   YU 7192  12:30 AM
4    Rockford (RFD)                   UPS    5X 108  12:48 AM

  Terminal   Status Arrival_time  A_day Scheduled    S_day
0      NaN   Landed          NaN    NaN       NaN      NaN
1      NaN  Unknown          NaN    NaN       NaN      NaN
2        B   Landed      1:48 PM    NaN   1:35 PM      NaN
3        B   Landed      1:48 PM    NaN   1:35 PM      NaN
4      NaN  Unknown          NaN    NaN       NaN      NaN

In [3]: ewr.shape
Out[3]: (1555, 10)
```

There are 1555 flights listed and 10 column headings: `Destination`, `Airline`, `Flight`, `Departure`, `Terminal`, `Status`, `Arrival_time`, `A_day`, `Scheduled`, and `S_day`. We will explain the headings as we go.

Let's get familiar with the ewr DataFrame. You might wonder what the possibilities are for the status of a flight. You can find out, as well as get some additional information, using the `value_counts()` method.

```
In [4]: ewr['Status'].value_counts()
Out[4]:
Landed - On-time       757
Landed - Delayed       720
Canceled                41
Landed                  18
En Route - Delayed      10
Unknown                  4
Scheduled - Delayed      2
En Route - On-time       1
```

```
En Route                   1
Diverted                   1
Name: Status, dtype: int64

In [5]: ewr['Status'].value_counts().sum()
Out[5]: 1555
```

The `value_counts()` method is quite useful. It finds all the unique entries in a Series (or DataFrame column) and reports the number of times each entry appears. We also checked to confirm that the counts for all the categories summed to the total number of entries.

Newark Airport has three terminals: A, B, and C. Let's find out how many departures there were from each terminal.

```
In [6]: ewr['Terminal'].value_counts()
Out[6]:
C    826
A    471
B    191
```

10.7.1 The groupby method

Now suppose we would like to know the status of each flight broken down by terminal. For this, we need a more sophisticated tool: groupby. Here is how it works:

```
In [7]: ewr['Status'].groupby(ewr['Terminal']).value_counts()
Out[7]:
Terminal  Status
A         Landed - On-time       229
          Landed - Delayed       218
          Canceled                21
          Landed                   3
B         Landed - On-time       104
          Landed - Delayed        70
          En Route - Delayed       6
          Canceled                 4
          Landed                   4
          Scheduled - Delayed      2
          En Route - On-time       1
C         Landed - Delayed       413
          Landed - On-time       395
          Canceled                14
          En Route - Delayed       4
Name: Status, dtype: int64
```

In this case, we want to know the status of each flight, so `ewr['Status']` comes first in our command above. Next, we want the status broken down by terminal, so we add the method `groupby` with the argument `ewr['Terminal']`. Finally, we want to know how many flights fall into each category so we add the method `value_counts()`.

Alternatively, we could have written

```
In [8]: ewr_statterm = ewr['Status'].groupby(ewr['Terminal'])
```

which creates a `groupby` object that we can subsequently process. For example, we can get the total number of flights from each terminal:

```
In [9]: ewr_statterm.count()
Out[9]:
Terminal
Terminal
A      471
B      191
C      826
Name: Status, dtype: int64
```

Or we can write `ewr_statterm.value_counts()`, which gives the same output as `Out [7]:` above.

10.7.2 Iterating over groups

Sometimes it is useful to iterate over groups to perform a calculation. For example, suppose that for each airline, we want to determine what fraction of the flights arriving at their destination arrived on time.

The information about on-time arrivals is contained in the `Status` column of the `ewr` DataFrame. It has, amongst other things, entries `Landed - On-time` and `Landed - Delayed`. We will want to use these entries to perform the calculation.

To do this, we use a `for` loop with the following construction:

```
for name, group in grouped:
```

where `grouped` is a `groupby` object, and `name` and `group` are the individual names and groups within the `groupby` object that are looped over.

To perform our calculation, we need to iterate over each airline, so our `groupby` object should group by `ewr['Airline']`. Before actually doing the calculations, however, we illustrate how the loop works for our `groupby` object with a little demonstration program. In this program, the loop doesn't do any calculations; it simply prints out the `name` and `group` for each iteration with the following code:

Code: chapter10/programs/ewrGroupbyElements.py

```
1  import pandas as pd
```

Data Manipulation and Analysis: Pandas 275

```
ewr = pd.read_csv('ewrFlights20180516.csv')

for airln, grp in ewr.groupby(ewr['Airline']):
    print('\nairln = {}: \ngrp:'.format(airln))
    print(grp)
```

The output of this program is:

```
airln = ANA:
grp:
          Destination Airline    Flight Departure Terminal
134   San Francisco (SFO)    ANA  NH 7007   7:00 AM        C
189     Los Angeles (LAX)    ANA  NH 7229   7:59 AM        C
303         Chicago (ORD)    ANA  NH 7469   8:59 AM        C
438           Tokyo (NRT)    ANA  NH 6453  11:00 AM        C
562         Chicago (ORD)    ANA  NH 7569   1:20 PM        C
1140    Los Angeles (LAX)    ANA  NH 7235   6:43 PM        C
1533      Sao Paulo (GRU)    ANA  NH 7214  10:05 PM        C

               Status Arrival_time  A_day Scheduled  S_day
134   Landed - Delayed     11:13 AM    NaN  10:18 AM    NaN
189   Landed - On-time     10:57 AM    NaN  11:05 AM    NaN
303   Landed - On-time     10:39 AM    NaN  10:25 AM    NaN
438   Landed - On-time      1:20 PM    NaN   1:55 PM    NaN
562   Landed - Delayed      3:16 PM    NaN   2:44 PM    NaN
1140  Landed - Delayed      9:54 PM    NaN   9:41 PM    NaN
1533  Landed - Delayed     10:06 AM    1.0   8:50 AM    1.0

airln = AVIANCA:
grp:
          Destination  Airline    Flight Departure Terminal
81         Dulles (IAD)  AVIANCA  AV 2135   6:05 AM        A
367        Dulles (IAD)  AVIANCA  AV 2233  10:00 AM        A
422         Miami (MIA)  AVIANCA  AV 2002  10:44 AM        C
805   San Salvador (SAL) AVIANCA  AV  399   3:55 PM        B
890         Bogota (BOG) AVIANCA  AV 2245   4:45 PM        C

                Status Arrival_time  A_day Scheduled  S_day
81    Landed - On-time      7:17 AM    NaN   7:25 AM    NaN
367   Landed - On-time     11:10 AM    NaN  11:20 AM    NaN
422   Landed - On-time      1:30 PM    NaN   1:46 PM    NaN
805  Scheduled - Delayed         NaN    NaN   7:05 PM    NaN
890   Landed - Delayed     12:42 AM    1.0   9:35 PM    NaN
  .
  .
  .
```

By examining this output, the form of the data structures being looped over should become clear to you.

Now let's do our calculation. To keep things manageable, let's say that we only care about those airlines that landed 12 or more flights. Grouping the data by airline, we perform the calculation using a `for` loop, accumulating the results about on-time and late flights in a list of lists, which we convert to a DataFrame at the end of the calculations.

```
In [10]: ot = []    # create an empty list to accumulate results
```

```
In [11]: for airln, grp in ewr.groupby(ewr['Airline']):
    ...:     ontime = grp.Status[grp.Status ==
                              'Landed - On-time'].count()
    ...:     delayd = grp.Status[grp.Status ==
                              'Landed - Delayed'].count()
    ...:     totl = ontime+delayd
    ...:     if totl >= 12:
    ...:         ot.append([airln, totl, ontime/totl])
```

The output of the code is a list called `ot`. We convert it to a DataFrame using the Pandas function `DataFrame.from_records`.

```
In [12]: t = pd.DataFrame.from_records(ot, columns=['Airline',
    ...:                   'Flights Landed', 'On-time fraction'])
```

We choose to print out the results sorted by on-time fraction, from largest to smallest.

```
In [13]: t.sort_values(by='On-time fraction', ascending=False)
Out[13]:
              Airline  Flights Landed  On-time fraction
0          Air Canada             129          0.472868
1           Air China              24          0.750000
2     Air New Zealand              34          0.617647
3     Alaska Airlines              20          0.500000
4   American Airlines              27          0.592593
5            Austrian              28          0.428571
6    Brussels Airlines              21          0.523810
7           CommutAir              47          0.531915
8        Copa Airlines              12          0.333333
9      Delta Air Lines              33          0.606061
10          ExpressJet              64          0.531250
11               FedEx              27          0.555556
12      JetBlue Airways             24          0.416667
13            Lufthansa            119          0.403361
14     Republic Airlines             66          0.606061
15                  SAS             94          0.414894
16                SWISS             20          0.450000
17   Southwest Airlines             18          0.500000
18         TAP Portugal             38          0.315789
```

Data Manipulation and Analysis: Pandas

```
19      United Airlines             417       0.529976
20      Virgin Atlantic              26       0.615385
```

10.7.3 Reformatting DataFrames

We often create DataFrames that need to be reformatted for processing. The range of reformatting issues that one can run across is enormous, so we can't even hope to cover all the eventualities. But we can illustrate a few to give you a sense of how this works.

In this example, we would like to perform an analysis of on-time arrivals. To do so, we will need to work with the Departure, Arrival_time, and Scheduled columns. Let's take a look at them.

```
In [14]: ewr[['Departure', 'Arrival_time', 'Scheduled']].head()
Out[14]:
  Departure Arrival_time Scheduled
0  12:09 AM         NaN       NaN
1  12:10 AM         NaN       NaN
2  12:30 AM     1:48 PM   1:35 PM
3  12:30 AM     1:48 PM   1:35 PM
4  12:48 AM         NaN       NaN
```

Some of the times are missing, represented by NaN in a DataFrame, but these are generally not much of a concern as Pandas handles them in an orderly manner. More worrisome are the times, which do not contain a date. This can be a problem if a flight departs on one day but arrives the next day. The ewr columns S_day and A_day for a particular row have entries of 1, respectively, if the flight is scheduled to arrive or if it actually arrives on the next day.

Before addressing these problems, let's examine the data types of the different columns of the ewr DataFrame.

```
In [15]: ewr.dtypes
Out[15]:
Destination     object
Airline         object
Flight          object
Departure       object
Terminal        object
Status          object
Arrival_time    object
A_day          float64
Scheduled       object
S_day          float64
dtype: object
```

We note that Departure, Arrival_time, and Scheduled are not formatted as datetime objects. To convert them to datetime objects, we use Pandas' `apply` method, which applies a function to a column (the default) or a row (by setting the keyword `axis=0`) of a DataFrame. Here, we use the Pandas function `pd.to_datetime`.

```
In [16]: ewr[['Departure', 'Arrival_time', 'Scheduled']] = \
    ...:     ewr[['Departure','Arrival_time', 'Scheduled']] \
    ...:     .apply(pd.to_datetime)

In [17]: ewr.dtypes
Out [17]: ewr.dtypes
Destination             object
Airline                 object
Flight                  object
Departure       datetime64[ns]
Terminal                object
Status                  object
Arrival_time    datetime64[ns]
A_day                  float64
Scheduled       datetime64[ns]
S_day                  float64
dtype: object
```

Next we set the dates. First, we use the datetime `replace` method to reset the year, month, and day of all dates to the departure date for all the flights: 2018-05-16.

```
In [18]: for s in ['Departure', 'Arrival_time', 'Scheduled']:
    ...:     ewr[s] = ewr[s].apply(lambda dt:
    ...:         dt.replace(year=2018, month=5, day=16))
```

Finally, we add a day to those dates in the Scheduled and Arrival_time columns that have a 1 in the corresponding S_day and A_day columns with this code snippet.

Code: chapter10/programs/addDaySnippet.py

```python
from datetime import timedelta
i_A = ewr.columns.get_loc("Arrival_time")   # i_A  = 6
i_Ap = ewr.columns.get_loc("A_day")         # i_Ap = 7
i_S = ewr.columns.get_loc("Scheduled")      # i_S  = 8
i_Sp = ewr.columns.get_loc("S_day")         # i_Sp = 9
for i in range(ewr.shape[0]):
    if ewr.iloc[i, i_Ap] >= 1:
        ewr.iloc[i, i_A] += timedelta(days=ewr.iloc[i, i_Ap])
    if ewr.iloc[i, i_Sp] >= 1:
        ewr.iloc[i, i_S] += timedelta(days=ewr.iloc[i, i_Sp])
```

Data Manipulation and Analysis: Pandas   279

After running this, the datetime stamps are correct for all the datetime entries, which we check by printing out some times for flights that departed late in the day.

```
In [19]: ewr[['Departure', 'Arrival_time', 'Scheduled']][-45:-40]
Out[19]:
              Departure           Arrival_time              Scheduled
1510 2018-05-16 21:55:00  2018-05-17 11:25:00  2018-05-17 11:40:00
1511 2018-05-16 21:57:00  2018-05-17 00:53:00  2018-05-16 23:05:00
1512 2018-05-16 21:57:00  2018-05-17 00:53:00  2018-05-16 23:05:00
1513 2018-05-16 21:59:00  2018-05-16 23:29:00  2018-05-16 23:42:00
1514 2018-05-16 21:59:00  2018-05-16 23:29:00  2018-05-16 23:42:00
```

Let's calculate the difference between the actual Arrival_time and the Scheduled arrival time in minutes.

```
In [20]: late = (ewr['Arrival_time']
                 - ewr['Scheduled']).dt.total_seconds()/60

In [21]: late.hist(bins=range(-50, 300, 10))
Out[21]: <matplotlib.axes._subplots.AxesSubplot at 0x1245b39b0>
```

Note that instead of setting the number of bins, as we have done previously, we specify the widths of the bins and their precise placement using the range function.

Let's go ahead and add axis labels to our plot, which is displayed in Fig. 10.7.

```
In [22]: xlabel('minutes late')
```

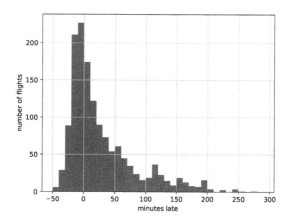

Figure 10.7   Histogram of late arrival times.

```
Out[22]: Text(0.5,23.5222,'minutes late')

In [23]: ylabel('number of flights')
Out[23]: Text(38.2222,0.5,'number of flights')
```

10.7.4 Custom aggregation of DataFrames

Pandas has a number of built-in functions and methods for extracting useful information from Pandas Series and DataFrames, some of which are listed in Table 10.3. These can be thought of as aggregation functions because they aggregate a set of data into some scalar quantity, like the minimum or maximum of a data set.

In addition to these built-in methods, Pandas has a method `agg` that allows you to implement your own aggregation functions. Suppose, for example, we would like to characterize the distribution of late arrival times plotted in Fig. 10.7. We could use the `std` method to characterize the width of distribution, but that would miss the fact that the distribution is obviously wider on the late (positive) side than it is on the early (negative) side.

To take this asymmetry into account, we devise our own function `siglohi` that calculates two one-sided measures of the width of the distribution.

Code: chapter10/programs/siglohi.py

```
1  def siglohi(x, x0=0, n=2):
2      xplus = x[x > x0] - x0
3      xminus = x0 - x[x < x0]
4      sigplus = ((xplus**n).mean())**(1/n)
5      sigminus = ((xminus**n).mean())**(1/n)
6      return sigminus, sigplus
```

By default, the function calculates the square root of the negative and positive second moments about zero. Using its optional keyword arguments, it can calculate the nth root of the nth moment and can center the calculation of the moments around any value, not just zero.

We demonstrate how `siglohi` works on the Series `late` of the distribution of flight arrival times that we developed in the previous section. We use the Pandas `agg` method on the Series `late` with our function `siglohi` as the argument of `agg`.

```
In [24]: late.agg(siglohi)
Out[24]: (16.613569037283458, 78.8155571711229)
```

Data Manipulation and Analysis: Pandas

As expected, the width is much smaller on the early (nagative) side than it is on the late (positive) side.

The optional keyword arguments of `siglohi` are passed in the usual way (see §7.1.6). For example, to calculate the cube root of the third moment, we set the optional argument `n` equal to 3 as follows:

```
In [25]: late.agg(siglohi, *(0, 0, 3))
Out[25]: (18.936255193664774, 96.23261210488258)
```

Note that there are *three*, not two, optional arguments. The first is the `axis`, which is an optional argument (the only one) for the `agg` method, and the second and third are `x0` and `n`, the optional arguments of `siglohi`. Alternatively, we can call `agg` with the `axis` argument of `agg` set to `zero` as a positional argument as follows.

```
In [26]: late.agg(siglohi, 0, *(0, 3))
Out[26]: (18.936255193664774, 96.23261210488258)
```

Either way, the result is the same.

Finally, we note that `siglohi` can be used on `late` as a function in the usual way with `late` as an explicit argument of `siglohi`.

```
In [27]: siglohi(late, n=3)
Out[27]: (18.936255193664774, 96.23261210488258)
```

10.8 Exercises

1. Read the planetary data in the text file `planetData.txt` into a Pandas DataFrame and perform the following tasks.

    (a) Based on the data read in from the file `planetData.txt`, find the average density of each planet relative to that of the Earth and add the results as a column in your DataFrame.

    (b) Print out your DataFrame sorted from the largest- to smallest-diameter planet.

    (c) Make a list of planets that have masses greater than that of Earth, sorted from least to most massive planet.

2. Starting from the program `urlRead.py` on page 261, write a program that compares the fluctuations of all the currencies relative to the US dollar (or some other currency of your choosing). Your code should find the average a, maximum m, and standard deviation s of the value of each currency relative to the US dollar over

the period of time starting from the first business day of 2017, January 3rd. Then create a DataFrame with columns that list mx = *m/a* and sd = *s/a* along with the name of each currency, as shown in the listing below. The DataFrame should be sorted from the largest mx to the smallest.

```
              mx      sd      description
id
FXZARCAD    1.119   0.0529   South African rand
FXTRYCAD    1.090   0.0485          Turkish lira
FXGBPCAD    1.088   0.0435   UK pound sterling
FXMYRCAD    1.084   0.0479   Malaysian ringgit
   .
   .
   .
```

3. Starting from the program urlRead.py on page 261, extend the code to make a plot like the one below. The three traces in each of the two plots give, respectively, the daily exchange rate, the daily exchange rate as a centered running average over 10 (business) days, and over 30 days. Look up the Pandas rolling averages

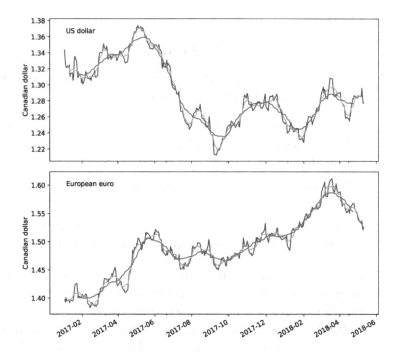

routine `pd.Series.rolling`, which you should find useful in making this plot. Write your program in such a way that you can switch the currency plotted by changing a single variable.

4. Go to the website https://www.ncdc.noaa.gov/cdo-web/search and make a request to download weather data for some place that interests you. I requested weather data as a csv file for Central Park in New York City (zip code 10023), as it dates back from the 19th century, although I chose to receive data dating from January 1, 1900.

   (a) Read the weather data you download into a DataFrame, tak-

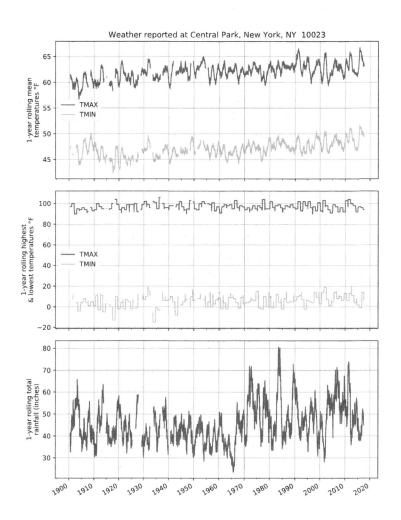

ing care to make sure the date is formatted as a datetime object and set to be the DataFrame index. Print out a table with the date as the first column and the daily precipitation, maximum temperature, and minimum temperature for one month of your choosing. The headings for those data are PRCP, TMIN, and TMAX, respectively.

(b) Get a list of the dates when more than 5 inches of rain fell and the rainfall on those dates. If this does not yield any results for your data, reduce the number of inches until you get a few dates.

(c) Make a plot like the one on the previous page from the data set you downloaded. The top graph plots the 1-year running averages (centered) of the daily high and low temperatures. The middle graph plots the running 1-year high and low temperatures, which are remarkably stable around 0°F and 100°F. The bottom graph plots the 1-year running total rainfall. Write the program that makes these graphs so that the 1-year running quantities can be changed to 2, 3, or some other number of years. Look up the Pandas rolling averages routine `pd.Series.rolling`, which you should find useful in making this plot.

5. In this problem you characterize the 46 pairs of chromosomes (22 pairs of autosomes and one pair of sex chromosomes) of the human genome working with a csv file downloaded from the Ensembl genome database at http://useast.ensembl.org/. The data file is called `humanGenes.csv`. Read this data file into a Pandas DataFrame and perform the following analyses.

    (a) Compute and print out the number of genes listed for the human genome (there is one per row).

    (b) Compute and print out the minimum, maximum, average, and median number of known isoforms per gene (consider the `transcript_count` column as a Series).

    (c) Plot a histogram of the number of known isoforms per gene. As these numbers vary over a wide range, use a logarithmic y-axis, as shown in the upper right plot in the figure below.

    (d) Compute and print out the number of different gene types.

(e) Compute and print out the total number of genes and the number of genes for each `gene_type`. Make a horizontal bar graph that shows the number of genes for each type associated with each gene in decreasing order of gene type.

(f) Compute and print out the number of different chromosomes.

(g) Compute and print out the number of genes for each chromosome. Make a vertical bar plot of the number of genes for each chromosome in decreasing order.

(h) Compute and print out the percentage of genes located on the + strand for each chromosome.

(i) Compute and print out the average number of transcripts associated with each gene type.

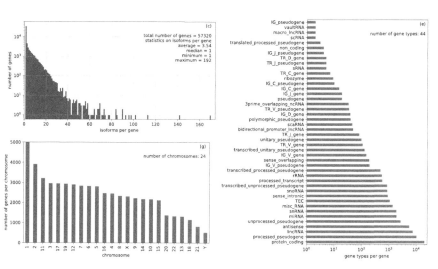

CHAPTER 11

Animation

> In this chapter you learn how to use matplotlib's Animation package. You learn how to **animate a sequence of images** to make a video, and then how add text and other features to your videos. You learn how to **animate functions**. You also learn how to combine movies and animated plots side-by-side. You learn how to animate a fixed number of frames or, alternatively, how to animate until some condition is met.

It's often not enough to see our data plotted, we want to see it move! Simulations, dynamical systems, wave propagation, explosions—they all involve time evolution. Moreover, the human brain is particularly well-adapted to extract and understand spatial information in motion. For all these reasons, we want to animate our representations of information.

While not strictly necessary, we use the *Animation* library of matplotlib to make animations as it is particularly well suited for our needs: animating functions, data, and images.

11.1 Animating a Sequence of Images

One of the most basic animation tasks is to make a movie from a sequence of images stored in a set of image files. If the size and number of images are not too large, you can simply read all the images into your program (*i.e.*, into memory) and then use the function ArtistAnimation of matplotlib's Animation class to play a movie. You can also save the movie you make to an external file using the save function from the Animation module.

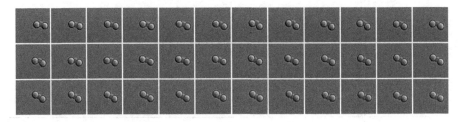

Figure 11.1 Partial sequence of images for animation.

11.1.1 Simple image sequence

First, we make a video from a sequence of images. Once this is done, we will show you how to add text and other animated features to your movie.

We will make a movie from a sequence of images of micrometer-size particles suspended in water that are undergoing Brownian motion. A selection of the sequence of images is shown in Fig. 11.1.

The names of the sequence of image files to be animated should consist of a base alphanumeric string—any legal filename—followed by an n-digit integer, including leading zeros, so that every file has a name with the same number of characters. The images are to be animated from the smallest to the largest numbers. As an example, suppose we want to animate a sequence of 100 image files named s000.png, s001.png, s002.png, ..., s099.png.

Here is our program. Below we explain how it works.

Code: chapter11/programs/movieFromImages.py

```
import matplotlib.pyplot as plt
import matplotlib.animation as anim
from PIL import Image
from glob import glob

fig, ax = plt.subplots(figsize=(3.6, 3.5))
fig.subplots_adjust(bottom=0, top=1, left=0, right=1)
ax.axis('off')

ims = []
for fname in sorted(glob('pacb/s0*.png')):
    # print(fname)   # uncomment to follow loading of images
    im = ax.imshow(Image.open(fname), animated=True)
    ims.append([im])

ani = anim.ArtistAnimation(fig, artists=ims, interval=33,
```

```
17                     repeat=False)
18  # Uncomment to save as mp4 movie file.  Need ffmpeg.
19  # ani.save('pacb.mp4', writer='ffmpeg')   # ffmpeg
20
21  fig.show()
```

The animation of the file sequence we read in is done by a function named `ArtistAnimation` which is part of the `matplotlib.animation` library. It's called at lines 16–17 and, in this example, has four arguments.

The first argument is the name of the figure window, in this case `fig`, where the animation will be rendered.

The second argument, with the keyword `artists`, must be a list of lists (or tuples) that contain the images to be animated. We explain below how such a list is put together.

The third argument, `interval`, specifies the time in milliseconds between successive frames in the animation. In this example, it's 30 ms, which corresponds to 1000/30 = 33.3 frames per second.

The fourth argument, `repeat`, tells the animation just to play through one time when it's set to `False`, rather than repeating in a loop over and over again.

It is important that the `ArtistAnimation` call (line 16) be assigned a variable name, which here is `ani`. For one thing, it's needed in line 18 (if it's uncommented) where the animation is saved to a movie file. But even if the movie is not saved, it is important to assign the `anim.ArtistAnimation` call to a variable, or it will be quite literally thrown out with the garbage—that is, the animation will be deleted before it can display the sequence of images. So don't forget to give any `ArtistAnimation` call a name!

Aside from calling the function `ArtistAnimation`, the main tasks for the program are to set up the figure window and then assemble the list `ims` that contains the images to be rendered for the animation.

Lines 6–8 set up the figure window. The argument `figsize` is set to have the same aspect ratio as the frames of the movie we want to animate. Then the function `subplots_adjust` is set so that frames take up the entire figure window. In line 8, we turn off all the axes labels, as we do not want them for our animation.

In line 10, we create an empty list, `ims`, that will contain the images to be animated.

The `for` loop starting at line 11 reads in an image frame from a sequence of image files, formats it for animation, and then adds it to the list of images to be animated.

To read in the names of our data files, we use the function `glob` from the module of the same name. The function `glob` returns a list of paths on your computer matching a pathname pattern. The asterisk symbol `*` acts as a wildcard for any symbol. Typing `glob('pacb/s*.png')` returns the list of all files on my computer matching this pattern, which turns out to be a sequence of 100 image files named `s000.png`, `s002.png`, `s003.png}`, `\ldots`, `\texttt{s099.png` that are located in the `pacb` subdirectory of the directory where our program is stored. To ensure that the list of filenames is read in numerical order, we use the Python function `sorted` in line 11 (which may not be necessary if the timestamps on the files are in the correct order). We can restrict the wildcard using square brackets with entries that specify which characters must appear at least once in the wildcard string. For example `glob('pacb/s00*[0-2].png')` returns the list `['pacb/s000.png'`, `'pacb/s001.png'`, `'pacb/s002.png']`. You can experiment on your own to get a clearer idea about how `glob()` works with different wildcards. Uncommenting line 12 prints out the names of the data files read and parsed by `glob()`, which can serve as a check that `glob()` is working as expected.

Two functions are used in line 13: `PIL.Image.open()` from the Python Image Library (PIL) reads an image from a file into a NumPy array; `imshow()` from the matplotlib library displays an image, stored as a NumPy array, on the figure axes. The image is not displayed right away, but is stored with the name `im`. Note that in each iteration of the `for` loop, we are reading in one frame from the sequence of frames that make up the animation clip we are putting together.

In the final line of the `for` loop, we append `[im]` to the list `ims` that we defined in line 10 just before the `for` loop. Note that `im` is entered with square brackets around it so that `[im]` is a one-item list. Thus it is added as a list to the list `ims`, so that `ims` is a list of lists. This is the format that the function `ArtistAnimation` needs for the `artists` argument.

Animation

Finally, we save the movie[1] as an mp4 movie so that it can be played independently of the program that produced it (and without running Python). Alternatively, changing the file name to `'pacb.avi'` saves the movie as an avi file. The mp4 and avi movies and the Python code above that produced them are available at https://github.com/djpine/python-scieng-public.

As an alternative to the program provided and discussed above, we offer one that is self-contained so that you do not need to load a sequence of images from files. Instead, this program makes the images on the fly, purely for demonstration purposes. The program is adapted from an example provided on the matplotlib web site.[2]

Code: chapter11/programs/movieFromImagesAlt.py

```
import numpy as np
import matplotlib.pyplot as plt
import matplotlib.animation as anim


def f(x, y):
    return np.sin(x) + np.cos(y)


x = np.linspace(0, 2 * np.pi, 120)
y = np.linspace(0, 2 * np.pi, 120).reshape(-1, 1)

fig, ax = plt.subplots(figsize=(3.5, 3.5))
fig.subplots_adjust(bottom=0, top=1, left=0, right=1)
ax.axis('off')
ims = []
for i in range(120):
    x += np.pi / 20.
    y += np.pi / 20.
    im = ax.imshow(f(x, y), cmap=plt.get_cmap('plasma'),
                   animated=True)
    ims.append([im])

ani = anim.ArtistAnimation(fig, artists=ims, interval=10,
                           repeat_delay=0)
# Uncomment to save as mp4 movie file. Need ffmpeg.
# ani.save('sncs2d.mp4', writer='ffmpeg')  # need ffmpeg !)
```

---

[1] To save a movie to your computer, you need to install a third-party MovieWriter that matplotlib recognizes, such as FFmpeg. See §A.3 for instructions on how to download and install FFmpeg. Alternatively, you can comment out the `ani.save` call (line 26) so that the program runs without saving the movie.

[2] See https://matplotlib.org/examples/animation/dynamic_image2.html.

```
28
29  fig.show()
```

The 2D NumPy array is created with `f(x, y)` in lines 20–21, in place of reading in image files from disk. The only other notable difference is that here we let the animation repeat over and over. We set the delay between repetitions to be 0 ms so that the animation appears as an endless repeating clip without interruption.

11.1.2 Annotating and embellishing videos

It is often useful to add dynamic text or to highlight various features in a video. In the sequence of images animated in the program `movieFromImages.py` (see page 289), there are two outer particles that rotate around a central particle, forming a kind of ball-and-socket joint. We would like to highlight the angle that the joint forms and display its value in degrees as the system evolves over time. We can do this by adding some matplotlib Artists to each frame. Figure 11.2 shows one frame of what our program will eventually display.

To start, we need data that gives the positions of the three particles as a function of time. These data are provided in an Excel spreadsheet called trajectories.xlsx. The data are read into the program in line 20.

Next, we construct a list, `ims`, that will contain a set of lists that

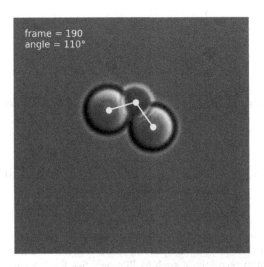

Figure 11.2   Annotated frame highlighting particle positions and displaying angle.

Animation

the animation routine will display. In a previous example each element of the list `ims` was a one-item list `[im]` of png images, (see line 14 in `movieFromImages.py` listed on page 289). Adding dynamic text and other features, this one-item list becomes a three-item list in the program below.

Code: chapter11/programs/movieFromImagesAnnotated.py

```
1  import numpy as np
2  import matplotlib.pyplot as plt
3  import matplotlib.animation as animation
4  import pandas as pd
5  from glob import glob
6  from PIL import Image
7
8
9  def angle(x, y):
10     a = np.array([x[0]-x[1], y[0]-y[1]])
11     b = np.array([x[2]-x[1], y[2]-y[1]])
12     cs = np.dot(a, b)/(np.linalg.norm(a)*np.linalg.norm(b))
13     if cs > 1.0:
14         cs = 1.0
15     elif cs < -1.0:
16         cs = -1.0
17     return np.rad2deg(np.arccos(cs))
18
19
20 r = pd.read_excel('trajectories.xlsx', usecols='A:F')
21
22 fig, ax = plt.subplots(figsize=(3.6, 3.5))
23 fig.subplots_adjust(bottom=0, top=1, left=0, right=1)
24 ax.axis('off')
25
26 ims = []
27 angles = []
28 for i, fname in enumerate(sorted(glob('pacb/s[0-2]*.png'))):
29     # print(fname)   # uncomment to follow loading of images
30     im = ax.imshow(Image.open(fname), animated=True)
31     # Make 3 solid points connect by two bars
32     x = np.array([r['x1'][i], r['xc'][i], r['x2'][i]])
33     y = np.array([r['y1'][i], r['yc'][i], r['y2'][i]])
34     ima, = ax.plot(x, y, 'o-', color=[1, 1, 0.7])
35     # Get angle between bars & write on movie frames
36     theta = angle(x, y)
37     angles.append(theta)
38     imb = ax.text(0.05, 0.95, 'frame = {0:d}\nangle = {1:0.0f}\u00B0'
39                   .format(i, theta), va='top', ha='left',
40                   color=[1, 1, 0.7], transform=ax.transAxes)
41     ims.append([im, ima, imb])
```

```
42
43  ani = animation.ArtistAnimation(fig, artists=ims, interval=33,
44                                  repeat=False)
45  # Uncomment to save as mp4 movie file.  Need ffmpeg.
46  # ani.save('movieFromImagesAnnotated.mp4', writer='ffmpeg')
47  fig.show()
```

The first item of the list is `im`, a list of the same png images we used before. This element is created in line 30.

The second item in the list is `ima`, a line plot connecting the centers of the three particles, where each center is indicated by a circular data point. This Artist is created in line 34. Note that a comma is used in defining `ima` because the `plot` function creates a one-element list, and we want the element itself, not the list.

The third item in the list is `imb`, a text Artist that displays the frame number and the angle of the ball-and-socket joint, which is calculated by the function `angle`. This Artist is created in lines 38–40.

The three items become the elements of a list `[im, ima, imb]` that represents one frame of our video: a png file, a plot, and text. Each frame, `[im, ima, imb]`, becomes an element in the list `ims`, which represents all the frames of the entire video.

The function `ArtistAnimation` is called with essentially the same inputs that we used previously. This time we choose not to have the video loop but instead we have it stop after it plays through one time.

Finally, you may have noticed that in line 28 we changed the argument of `glob`. The `[0-2]` is a wildcard that specifies that only 0, 1, and 2 will be accepted as the first character in the file name. In this way, a movie from `000` to `299` is made.

11.2 Animating Functions

Suppose you would like to visualize the nonlinear single or double pendulums whose solutions we calculated in Chapter 9. While it might not seem obvious at first, the simplest way to do these kind of animations is with the *function animation* routine, called `FuncAnimation`, of matplotlib's mpl's `Animation` library. As its name implies, `FuncAnimation` can animate functions, but it turns out that animating functions encompasses a wide spectrum of animation tasks, more than you might have imagined.

11.2.1 Animating for a fixed number of frames

We start by writing a program to animate a propagating wave packet with an initial width a_0 that spreads with time. The equation for the wave packet is given by the real part of

$$u(x,t) = \frac{1}{\sqrt{\alpha+i\beta t}} e^{ik_0(x-v_p t)} e^{-(x-v_g t)^2/4(\alpha+i\beta t)},$$

where $i \equiv \sqrt{-1}$, $\alpha = a_0^2$, and $\beta = v_g/2k_0$. The phase and group velocities are v_p and v_g, respectively, and $k_0 = 2\pi/\lambda_0$, where λ_0 is the initial wavelength of the wave packet. Figure 11.3 shows the wave packet at a particular moment in time.

Here is the program for animating the propagation of the wave packet. Following the listing, we explain how it was designed and how it works.

Code: chapter11/programs/wavePacketSpreads.py

```
1  import numpy as np
2  import matplotlib.pyplot as plt
3  import matplotlib.animation as anim
4
5
6  def ww(x, t, k0, a0, vp, vg):
7      tc = a0*a0+1j*(0.5*vg/k0)*t
8      u = np.exp(1.0j*k0*(x-vp*t)-0.25*(x-vg*t)**2/tc)
9      return np.real(u/np.sqrt(tc))
10
11
12 wavelength = 1.0
13 a0 = 1.0
14 k0 = 2*np.pi/wavelength
15 vp, vg = 5.0, 10.0
16 period = wavelength/vp
```

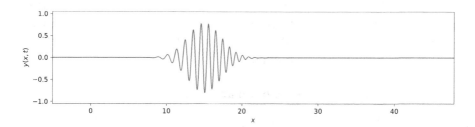

Figure 11.3 One frame from the propagating wave packet movie.

```
17  runtime = 40*period            # total time to follow wave
18  rundistance = 0.6*vg*runtime   # total distance to plot wave
19  dt = period/6.0                # time between frames
20  tsteps = int(runtime/dt)       # total number of times wave
21  # form is calculated
22  print('Frame time interval = {0:0.3g} ms'.format(1000*dt))
23  print('Frame rate = {0:0.3g} frames/s'.format(1.0/dt))
24
25  fig, ax = plt.subplots(figsize=(12, 3))
26  fig.subplots_adjust(bottom=0.2)   # allow room for axis label
27  x = np.arange(-5*a0, rundistance, wavelength/20.0)
28  line, = ax.plot(x, np.ma.array(x, mask=True), color='C0')
29  ax.set_xlabel(r'$x$')
30  ax.set_ylabel(r'$y(x,t)$')
31  ax.set_xlim(-5*a0, rundistance)
32  ax.set_ylim(-1.05, 1.05)
33
34
35  def animate(i):
36      t = float(i)*dt
37      line.set_ydata(ww(x, t, k0, a0, vp, vg))   # update y-data
38      return line,
39
40
41  ani = anim.FuncAnimation(fig, func=animate,
42                           frames=range(tsteps),
43                           interval=1000*dt, blit=True)
44  # Uncomment to save as mp4 movie file. Need ffmpeg.
45  ani.save('wavepacket.mp4', writer='ffmpeg')
46  fig.show()
```

After importing the relevant libraries, we define the wave packet in lines 6–9, using complex algebra in the calculation, but then taking only the real part at the end. The physical parameters defining various properties of the wave are initialized in lines 12–15. The range of times and distances over which the waveform will be calculated are determined in lines 17–19.

We calculate and print out the frame time interval and the frame rate in lines 22–23. There is not much point in animating at a frame rate higher than about 30 frames/s as that is about as fast as the human brain can perceive movement.

Setting up static elements of an animation

Next, in lines 25–32, we set up the *static* elements of the figure—everything about the figure *that does not change* during the animation. The *x*-array is defined in line 27, which indeed, remains fixed throughout the animation—it does not change with time. The distance between points along the x-axis is set to be small enough, $1/20^{\text{th}}$ of a wavelength, to make the waveform appear smooth.

Line 28 merits special attention. Here, the program is setting up a *container* for the animated plotting of the waveform. In so doing, we include the x-data but not the y-data. The reason is that the y-data will *change* as the wave propagates; at this point in our program, we are setting up the plot for only the *fixed* unchanging elements. Since we need a placeholder for the y-array, we simply use a fully masked x-array, which is guaranteed to have the right number of elements, but will not plot. You could simply put an appropriately initialized version of the y-array here without consequence, since the `plot` command should not be rendered until the `show()` command is called at the end of the routine. However, if you are running the program from IPython *and* you have interactive plotting turned on (`plt.ion()`), the `plot` command will be rendered immediately, which spoils the animation. Using a masked array for the y-data in the `plot` command avoids this problem so that the animation is rendered correctly irrespective of whether interactive mode is on or off.

Why is there a comma after an assigned variable name?

Notice that in line 28, there is a comma after the name `line`. This is important. To understand why, consider the following command issued from the IPython shell:

```
In [1]: plot([1, 2, 3, 2, 3, 4])
Out[1]: [<matplotlib.lines.Line2D at 0x181dca2e48>]
```

Notice that the `plot` command returns a one-item *list*, which is indicated by the square brackets around the matplotlib line object `<matplotlib.lines.Line2D at 0x181dca2e48>`. Writing `[line,] = plot(...)`, or equivalently writing `line, = plot(...)`, sets `line` equal to the first (and only) element of the list, which is the line object `<matplotlib.lines.Line2D at 0x181dca2e48>`, rather than the list. It is the line object that the `FuncAnimation` needs, not a list, so writing `line, = plot(...)` is the right thing to do. By the way, you could also write `line = plot(...)[0]`. That works too!

Lines 31–32 fix the limits of the x and y axes. Setting the plot limits to fixed values is generally recommended for an animation.

Animating a function

The animation is done in lines 41–43 by FuncAnimation from matplotlib's Animation package. The first argument is the name of the figure window, fig in this example, where the animation is to be rendered.

The second argument, func, specifies the name of the function, here animate, that updates each frame of the animation. We defer explaining how the function animate works until we've discussed the other arguments of FuncAninmation().

The sole input to the animate() function is the current value of the iterator provided by the third argument frame, which here is set equal to range(tsteps). The iterator serves two functions: (1) it provides data to func (here animate), in this case a single integer that is incremented by 1 each time a new frame is rendered; (2) it signals FuncAnimation to keep calling the routine animate until the iterator has run through all its values. At that point, FuncAnimation will restart the animation from the beginning unless an additional keyword argument, repeat, not used here, is set equal to False.

The interval argument sets the time in milliseconds between successive frames. Thus, the frame rate is 1000/interval frames/s.

The last argument, blit=True, turns on *blitting*. Blitting is the practice of redrawing only those elements in an animation that have changed from the previous frame, as opposed to redrawing the entire frame each time. This can save a great deal of time and allow an animation to run faster. In general we like to use blitting, so that an animation runs as close to the intended rate as possible.

We return now to explaining how the function animate updates each frame of the animation. In this example, updating means calculating the y-values of the wave packet for the next time step and providing those values to the 2D line object—the wave packet—that we named line in line 28. This is done using the set_ydata() function, which is attached via the *dot* syntax to the name line. The argument of set_ydata() is simply the array of updated y-values, which is calculated by the function ww().

FuncAnimation does the rest of the work, updating the animation frame by frame at the specified rate until the animation finishes. Once finished, the animation starts again since we did not set the keyword argument repeat=False.[3]

---

[3] If you consult the online documentation on FuncAnimation, you will see that

Animation

The next statement `ani.save('wavepacket.mp4')` saves the animation (one iteration only) in the current directory as an mp4 video that can be played with third-party applications on any platform (in principle).

Adding dynamic text

Our animated function is a function of three variables, x, y, and t, but we only show the values of x and y. We wish to remedy this problem by displaying the current value of time as dynamic *text* within the movie frame. To do this, we need to create a container for the text we wish to display, just as we did when we made the `line` container for the plot we wanted to display. We do this in the program below starting at line 28. Lines 1–28 are the same as in the previous program listing on page 295.

Code: chapter11/programs/wavePacketSpreadsEmb.py

```
28  line, = ax.plot(x, np.ma.array(x, mask=True), color='C0')
29  timeText = ax.text(0.9, 0.98, '', ha="left", va="top",
30                     transform=ax.transAxes)
31  timeString = "time = {0:0.2f}"
32  ax.text(0.9, 0.91, r'$v_p = {0:0.1f}$'.format(vp),
33          ha="left", va="top", transform=ax.transAxes)
34  ax.text(0.9, 0.84, r'$v_g = {0:0.1f}$'.format(vg),
35          ha="left", va="top", transform=ax.transAxes)
36  ax.set_xlabel(r'$x$')
37  ax.set_ylabel(r'$y(x,t)$')
38  ax.set_xlim(-5*a0, rundistance)
39  ax.set_ylim(-1.05, 1.05)
40
41
42  def animate(i):
43      t = float(i)*dt
44      line.set_ydata(ww(x, t, k0, a0, vp, vg))   # update y-data
45      timeText.set_text(timeString.format(t))
46      return line, timeText
47
48
49  ani = anim.FuncAnimation(fig, func=animate,
50                           frames=range(tsteps),
51                           interval=1000*dt, blit=True)
```

there is a keyword argument `init_func`, which the documentation states is used to draw a clear frame at the beginning of an animation. This function is superfluous as far as I can tell, so I don't use it, in spite of the fact that many web site tutorials suggest that it is necessary.

```
52  # Uncomment to save as mp4 movie file.  Need ffmpeg.
```

With the third argument an empty string, lines 29–30 create a blank text container that we will use in our animation to display the up-to-date time for the current frame. Lines 32–35 create two *static* texts that display the values of the phase and group velocities[4] underneath the dynamic time display. In each case, we include the keyword argument `transform=ax.transAxes` to specify the text in axis coordinates, where `0, 0` is lower-left and `1, 1` is upper-right; without this argument, data coordinates are used.

The `animate(i)` function includes lines for updating the text and line Artists in the animation. The rest of the program, with `FuncAnimation()` and `show()`, is unchanged from the previous version.

11.2.2 Animating until a condition is met

In the previous section, our animation of the propagating wave packet ran for a preset number of steps (`tsteps`). This is a sensible way to make an animation when you know or can calculate ahead of time how long you want the animation to run. In other cases, however, you may want the animation to run until some condition is met, and exactly how long this takes is not known ahead of time. This is generally the case when the animation involves some random process. A simple but powerful way to do this is to write a *generator* function, which is used as the `frames` keyword argument in `FuncAnimation`.

To illustrate this kind of animation, we introduce an algorithm known as *random organization*, which first appeared in the context of a physics problem.[5] In its simplest form, we consider a set of N spheres with a diameter of 1 that are randomly placed along the circumference of a circle of length $L > N$, as shown in Fig. 11.5. Nearby spheres overlap each other if the distance between their centers is less than 1. To aid visibility, spheres that overlap are colored differently from spheres that do not. The time evolution of the system proceeds as follows. Each time step, the subroutine `move` checks all N spheres to see which spheres, if any, overlap each other. Any sphere that is found to overlap with one or more spheres is then given a kick that moves it

---

[4]The phase velocity is the speed with which the crests in the wave packet moves; the group velocity is the overall speed of the wave packet. Don't worry if you're not familiar with these terms.

[5]Corté *et al.*, Nature Physics 4, 420–424, (2008).

a random distance between $-\epsilon$ and $+\epsilon$, where ϵ is typically about $1/4$ or less. Spheres that do not overlap with any of their neighbors do not move. That ends one time step. For any given time step, the number of spheres that overlap may increase, decrease, or remain the same. In the next time step, the process repeats. The algorithm continues as long as there are still spheres that overlap. In practice, it is found that all spheres eventually find a position at which they do not overlap with any other sphere if the number of spheres is not too high. For the conditions $L = 100$ and $\epsilon = 0.25$, the system eventually settles into a state where no spheres move if $N \leq 86$.

The random organization algorithm is implemented in the generator function move below. In this implementation, we use periodic boundary conditions, which is equivalent to bending the line on which the spheres move into a circle, so that spheres at the very end of the line interact with spheres at the beginning of the line.

The generator function move returns two arrays: x, which gives the updated positions of the N spheres as a floating point number be-

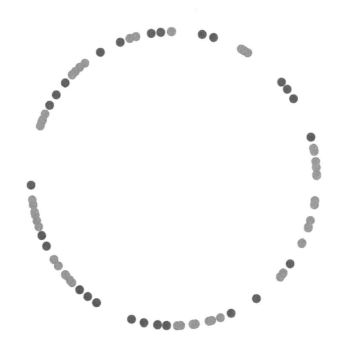

Figure 11.4   Random organization.

tween 0 and L, and changes, an integer array of length N where the ith entry is 1 if the ith sphere has moved in the most recent time step and 0 if it hasn't.

Code: chapter11/programs/randOrg.py

```
1  import numpy as np
2  import matplotlib.pyplot as plt
3  import matplotlib.animation as anim
4
5
6  def move(L, N, eps):                    # generator for updating
7      x = np.sort(L*np.random.rand(N))    # speeds up algorithm
8      moves = 1
9      while moves > 0:
10         changes = np.zeros(N, dtype="int")
11         xc = np.copy(x)
12         for i in range(N-1):
13             j = i + 1
14             while x[j]-x[i] < 1.0:
15                 rr = 2.0*(np.random.rand(2)-0.5)
16                 xc[i] += eps*rr[0]
17                 xc[j] += eps*rr[1]
18                 changes[i] = 1
19                 changes[j] = 1
20                 if j < N-1:
21                     j += 1
22                 else:
23                     break       # terminates while loop when j=N-1
24             if x[i] < 1.0:       # periodic boundary conditions
25                 k = -1
26                 while x[i] + L - x[k] < 1.0:
27                     rr = 2.0*(np.random.rand(2)-0.5)
28                     xc[i] += eps*rr[0]
29                     xc[k] += eps*rr[1]
30                     changes[i] = 1
31                     changes[k] = 1
32                     k -= 1
33         x = np.sort(xc % L)    # sort data for algorithm to work
34         moves = np.sum(changes)
35         yield x, changes
36
37
38 N, L, eps = 75, 100, 0.25    # inputs for algorithm
39
40 circumference = float(L)
41 radius = circumference/(2.0*np.pi)
42 R = radius*np.ones(N)
43
```

```
44  fig, ax = plt.subplots(figsize=(8, 8),
45                         subplot_kw=dict(polar=True))
46  pStill, = ax.plot(np.ma.array(R, mask=True), R,
47                    'o', ms=12, color='C0')
48  pActiv, = ax.plot(np.ma.array(R, mask=True), R,
49                    'o', ms=12, color='C1')
50  ax.set_rmax(1.1*radius)
51  ax.axis('off')
52
53
54  def updatePlot(mv):
55      x, changes = mv
56      angle = 2.0*np.pi*x/L
57      active = np.ma.masked_where(changes != 1, angle)
58      inactive = np.ma.masked_where(changes == 1, angle)
59      pStill.set_xdata(inactive)
60      pActiv.set_xdata(active)
61      return pStill, pActiv
62
63
64  ani = anim.FuncAnimation(fig=fig, func=updatePlot,
65                           frames=move(L, N, eps),
66                           interval=10, save_count=200,
67                           blit=True, repeat=False)
68  # Uncomment to save as mp4 movie file. Need ffmpeg.
69  # ani.save('randOrg.mp4', writer='ffmpeg', dpi=200)
70  fig.show()
```

The output of move(L, N, eps) provides the input to the function update(mv), which updates the animation. First, it unpacks mv in line 55, then converts the x array into angles for display purposes, and then creates two masked arrays, one for active and the other for inactive particles using the array changes that tracks which spheres moved in the most recent time step. These updated masked arrays, created in lines 57 and 58, are fed into the plots that were set up in lines 46–49, where the static data for the *y*-array—the unchanging radius of the polar plot—was already entered. The still and active data sets are updated and returned to FuncAnimation which plots the next frame, with the moving particles shown in orange (color='C1') and the stationary particles in blue (color='C0').

A key feature of this approach is the use of a generator function, here move. The function returns two arrays, x and changes, using a yield statement, which is what makes the function a generator. Note that move yields these two arrays *inside* a while loop. A key feature of a generator function is that it remembers its current state between calls. In particular, move remembers the value of the positions x of all the spheres, it remembers that it is in a while loop, and it remembers the variable move which keeps track of how many spheres were moved in the most recent time step. When move is zero, the while loop terminates, which signals FuncAnimation that the animation is finished.

In line 50, we use set_rmax to set the maximum radius of the polar plot. It is important to do this after the plot calls in lines 46 and 48, as they can reset the plot limits in unanticipated ways.

You can save the plot as an mp4 movie by uncommenting line 69. However, you need to have installed FFmpeg, as discussed in a footnote on page 291. Since, in this case, the length of the movie is not known *a priori*, FuncAnimation has a keyword argument save_count that you can set to limit the number of frames that are recorded in the movie. Its default value is 100, so if you want to record more frames than that, you need to include it in the FuncAnimation call and set it to some other value. If the keyword frames is an iterable that has a definite length (not the case here), it will override the save_count keyword argument value.

When running the program, be aware that the animation will not begin displaying until the movie is recorded for whatever number of frames you set, so if the number is large you may have to wait awhile before the animation appears.

Gilding the lily

As nice as this animation is, it would seem helpful to plot the number of active particles as a function of time to give a better sense of how the system evolves. Therefore, in the program below we add a plot inside the circular animation of the spheres. Aside from three additional lines, which we discuss below, the program below is the same as the program starting on page 302 up to the line ax.axis('off').

Code: chapter11/programs/randOrgLily.py

```
54  ax.axis('off')
55
56  gs = gridspec.GridSpec(3, 3, width_ratios=[1, 4, 1],
57                         height_ratios=[1, 2, 1])
58  ax2 = fig.add_subplot(gs[4], xlim=(0, 250), ylim=(0, L))
59  ax2.set_xlabel('time')
60  ax2.set_ylabel('active particles')
61  activity, = ax2.plot([], [], '-', color='C1')
62  tm, number_active = [], []
```

Animation

```
63
64
65  def updatePlot(mv):
66      t, moves, x, changes = mv
67      tm.append(t)
68      number_active.append(moves)
69      tmin, tmax = ax2.get_xlim()
70      if t > tmax:
71          ax2.set_xlim(0, 2*tmax)
72      angle = 2.0*np.pi*x/L
73      active = np.ma.masked_where(changes != 1, angle)
74      inactive = np.ma.masked_where(changes == 1, angle)
75      pStill.set_xdata(inactive)
76      pActiv.set_xdata(active)
77      activity.set_data(tm, number_active)
78      return pStill, pActiv, activity
79
80
81  ani = anim.FuncAnimation(fig=fig, func=updatePlot,
82                           frames=move(L, N, eps),
83                           interval=10, save_count=200,
84                           blit=False, repeat=False)
85  # Uncomment to save as mp4 movie file. Need ffmpeg.
86  # ani.save('randOrgLily.mp4', writer='ffmpeg', dpi=200)
87  fig.show()
```

Lines 56–57 set up the area for the additional plot using the `gridspec` module from the matplotlib library. To add this library, we include this line with the other import statements at the beginning of the program (the first of the three additional lines before line 56):

```
import matplotlib.gridspec as gridspec
```

Returning to line 56, the first two arguments of `GridSpec()` set up a 3×3 grid in the figure frame. You can set the relative widths of the columns and rows using the `width_ratios` and `height_ratios` keyword arguments. Then in line 58, we select grid rectangle number 4, which is at the center of a grid, by setting the first argument of `add_subplot` to `gs[4]`. The rectangles are numbered starting at zero from left to right and top to bottom. The other arguments fix the range of the x and y axes. The next few lines set up the plot of number or active particles—the *activity*—as well as lists for the time `t` and number of active particles `move`.

The only difference in the generator function `move` is that it returns two more variables than it did previously. Now the final line reads:

```
        yield t, moves, x, changes
```

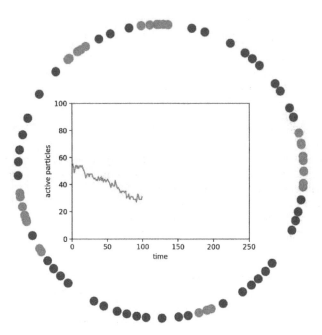

Figure 11.5 Random organization with a plot of the number of active particles *vs.* time.

We've inserted the current time `t` and the number of particles `moves` that moved in the most recent cycle. In the function `update`, the number `t` and the number `moves` are appended to the lists `tm` and `number_active` that keep a record of the number of active particles as a function of time. These data are then transmitted to the `activity` plot using the `set_data` function in line 77.

11.3 Combining Videos with Animated Functions

When presenting or analyzing scientific data, it's often useful to display a video with an animated plot that shows the evolution of some feature of the video. Consider, for example, the sequence of images we animated in §11.1.2. We would like to show the movie and next to it display the evolving distribution of the angles that are swept out by the ball-and-socket joint. Figure 11.6 shows what we are aiming for.

Animation

11.3.1 Using a single animation instance

Starting from the program on page 294, we modify line 22 by expanding the figure size and creating two Axes objects, one for the video and another for the plot of the distribution of angles. The distribution of angles is calculated using the NumPy histogram function on lines 38–39. Lines 40–43 set up the plot of the distribution, which is then appended to the list of Artists that are to be animated in line 44. The call to ArtistAnimation is the same as before.

Alternatively, we can make a histogram for the distribution of angles by commenting out lines 43–44 and uncommenting lines 46–47. matplotlib's bar function returns a special bar container that needs to be turned into a list for incorporation into the list of lists for the animation, which is done in line 47.

Code: chapter11/programs/movieFromImagesHistP.py

```
1  from glob import glob
2  import numpy as np
3  import matplotlib.pyplot as plt
4  import matplotlib.animation as anim
5  import pandas as pd
6  from PIL import Image
7
8
9  def angle(x, y):
10     a = np.array([x[0]-x[1], y[0]-y[1]])
11     b = np.array([x[2]-x[1], y[2]-y[1]])
12     cs = np.dot(a, b)/(np.linalg.norm(a)*np.linalg.norm(b))
```

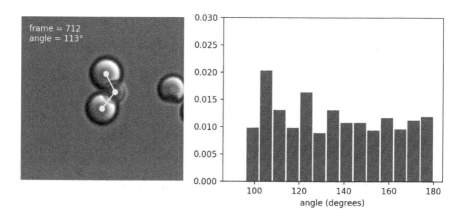

Figure 11.6   Movie with animated histogram.

```
13      if cs > 1.0:
14          cs = 1.0
15      elif cs < -1.0:
16          cs = -1.0
17      return np.rad2deg(np.arccos(cs))
18
19
20  r = pd.read_excel('trajectories.xlsx')
21
22  fig, (ax1, ax2) = plt.subplots(1, 2, figsize=(8, 3.5))
23  ax1.axis('off')
24
25  ims = []
26  angles = []
27  for i, fname in enumerate(sorted(glob('pacb/s0*.png'))):
28      # print(fname)   # uncomment to follow loading of image frames
29      im = ax1.imshow(Image.open(fname), animated=True)   # image
30      x = np.array([r['x1'][i], r['xc'][i], r['x2'][i]])   # 3 balls
31      y = np.array([r['y1'][i], r['yc'][i], r['y2'][i]])   # joined by
32      ima, = ax1.plot(x, y, 'o-', color=[1, 1, 0.7])       # 2 lines
33      theta = angle(x, y)
34      angles.append(theta)
35      imb = ax1.text(0.05, 0.95, 'frame = {0:d}\nangle = {1:0.0f}\u00
36                     .format(i, theta), va='top', ha='left',
37                     color=[1, 1, 0.7], transform=ax1.transAxes)
38      a, b = np.histogram(angles, bins=15, range=(90, 180),
39                          normed=True)
40      xx = 0.5*(b[:-1]+b[1:])
41      ax2.set_ylim(0, 0.03)
42      ax2.set_xlabel('angle (degrees)')
43      im2, = ax2.plot(xx, a, '-oC0')
44      ims.append([im, ima, imb, im2])
45      # plot histogram
46      # im2 = ax2.bar(xx, a, width=0.9*(b[1]-b[0]), color='C0')
47      # ims.append([im, ima, imb] + list(im2))
48      plt.tight_layout()
49
50  ani = anim.ArtistAnimation(fig, artists=ims, interval=33,
51                             repeat=False, blit=False)
52  # Uncomment to save as mp4 movie file. Need ffmpeg.
53  # ani.save('movieFromImagesHistP.mp4', writer='ffmpeg')
54  fig.show()
```

11.3.2 Combining multiple animation instances

Let's look at another example of a movie combined with a dynamic plot. Our previous example showed how to do this by rendering many

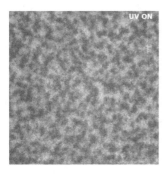
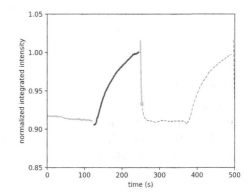

Figure 11.7 Movie with animated plot.

plotting elements using a single instance of ArtistAnimation. While ArtistAnimation is the natural choice for animating sequences of images, FuncAnimation is the more natural choice for animating a dynamic plot. So in this section, we use two different animation instances, one using ArtistAnimation to animate an image sequence and the other using FuncAnimation to animate the plot. We then show how to combine them into a single animation.

One frame of the result is shown in Fig. 11.7. As the movie progresses, the illumination switches from ultraviolet (uv) to blue, which is designated on the image in the upper right corner and reflected in a change in color from violet to blue in the trace on the left.

After reading the sequence of frames to make the movie, the program reads in data associated with each frame from a csv file.

The static part of the plot, which is not animated, is rendered first. Lines 23–28 set up containers for the animated lines and circle, which change color according to whether uv or blue light is used to illuminate the image sequence.

Before plotting the data, the uv and blue data are masked in lines 31–32 so that only one of the two traces is displayed at any given time.

The for loop starting at line 36 puts together the list ims of the frames to be animated. Each frame is itself a list of the separate elements to be rendered in each frame: an image and the UV ON/UV OFF text. The if-else block increases the brightness of the frames when uv light is off so that all the frames of the movie have nearly the same brightness. The loop is completed when the list of plot elements is appended to the ims.

Next, the routine `animate` is defined, which is called by the `FuncAnimation` routine to animate the line plot.

`ArtistAnimation` is called to animate the movie frames and `FuncAnimation` is called to animate the line plot. The second animation `ani2` is synchronized to the event clock of `ani1` using the keyword argument `event_source=ani1.event_source`. This assures that the two animations are updated by a single clock.

Finally, to save both animations to the same file, we set the keyword argument, which takes a list (or tuple), `extra_anim=(ani1, )` in our `ani2.save` call.

Code: chapter11/programs/movieSyncPlot1.py

```
1  import numpy as np
2  import matplotlib.pyplot as plt
3  import matplotlib.animation as anim
4  from PIL import Image, ImageEnhance
5  from glob import glob
6
7  framesDir = 'movieSyncFrames'      # movie frames directory
8  framesData = 'movieSyncData.csv'   # data file with intensities
9  time, uv, blue = np.loadtxt(framesData, skiprows=1,
10                             unpack=True, delimiter=',')
11
12 # Static parts of plot come first
13 fig, (ax1, ax2) = plt.subplots(1, 2, figsize=(9, 4))
14 fig.subplots_adjust(bottom=0.15, top=0.95, left=0, right=0.98)
15 ax1.axis('off')
16 ax2.set_xlim([0, time.max()])
17 ax2.set_ylim([0.85, 1.05])
18 ax2.plot(time, uv+blue, dashes=(5, 2), color='gray', lw=1)
19 ax2.set_xlabel('time (s)')
20 ax2.set_ylabel('normalized integrated intensity')
21 ax2.set_yticks([0.85, 0.9, 0.95, 1., 1.05])
22 # Set up plot containers for ax2
23 plotdotUV, = ax2.plot(np.nan, np.nan, 'o', color='violet',
24                       ms=6, alpha=0.7)
25 plotdotBlue, = ax2.plot(np.nan, np.nan, 'o', color='blue',
26                         ms=6, alpha=0.7)
27 plotlineB, = ax2.plot(np.nan, np.nan, '-', color='blue', lw=2)
28 plotlineU, = ax2.plot(np.nan, np.nan, '-', color='violet', lw=2)
29
30 # Mask data you do not want to plot
31 uvM = np.where(uv > 0.9, uv, np.nan)
32 blueM = np.where(blue > 0.9, blue, np.nan)
33
34 # Dynamic parts of plot come next
35 ims = []
```

Animation

```
36  for i, fname in enumerate(sorted(glob(framesDir+'/sp*.png'))):
37      # print(fname)   # uncomment to follow loading of image frames
38      if uv[i] >= blue[i]:
39          im = ax1.imshow(Image.open(fname), animated=True)
40          textUV = ax1.text(320, 20, 'UV ON', color='white',
41                            weight='bold')
42      else:
43          img0 = Image.open(fname)
44          # Increase brightness of uv-illuminated images
45          img0 = ImageEnhance.Brightness(img0).enhance(2.5)
46          im = ax1.imshow(img0, animated=True)
47          textUV = ax1.text(320, 20, 'UV OFF', color='yellow',
48                            weight='bold')
49      ims.append([im, textUV])
50
51
52  def animate(i):
53      plotdotUV.set_data(time[i], uvM[i])
54      plotdotBlue.set_data(time[i], blueM[i])
55      plotlineB.set_data(time[0:i], blueM[0:i])
56      plotlineU.set_data(time[0:i], uvM[0:i])
57      return plotdotUV, plotdotBlue, plotlineB, plotlineU
58
59
60  ani1 = anim.ArtistAnimation(fig, artists=ims, interval=33,
61                              repeat=False)
62  ani2 = anim.FuncAnimation(fig, func=animate,
63                            frames=range(time.size), interval=33,
64                            repeat=False, blit=False,
65                            event_source=ani1.event_source)
66  # Uncomment to save as mp4 movie file.  Need ffmpeg.
67  # ani2.save('movieSyncPlot1.mp4', extra_anim=(ani1, ),
68  #           writer='ffmpeg', dpi=200)
69  fig.show()
```

11.4 Exercises

1. Write a program to animate a 2-dimensional random walk for a fixed number of steps. Midway through the animation, your animation should look something like this.

   Start the random walk at $x = y = 0$. Show the leading edge of the random walk as a red circle and the rest of the walk as a line. Make sure the line extends from the starting point $(0,0)$ through to the red circle at the end. The x and y axes should span the same distance in an equal-aspect-ratio square plot.

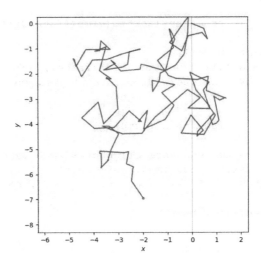

The following code gets you started by creating a 2D random walk of $N = 200$ steps.

Code: chapter11/exercises/Exercise01/diffusion.py

```
1  import numpy as np
2  import matplotlib.pyplot as plt
3  import matplotlib.animation as anim
4
5  N = 200
6  # Create arrays of random jump lengths in random directions
7  dr = np.random.random_sample(N-1)    # random number 0-1
8  angle = 2.0*np.pi*np.random.random_sample(N-1)
9  dx = dr*np.cos(angle)
10 dy = dr*np.sin(angle)
11 # Add up the random jumps to make a random walk
12 x = np.insert(np.cumsum(dx), 0, 0.)  # insert 0 as
13 y = np.insert(np.cumsum(dy), 0, 0.)  # first point
```

2. Embellish the animation of Exercise 1 by adding x vs. t and y vs. t panels along with animated text that gives the number of steps so far. Take t to be equal to the running number of steps. The result should look something like this:

3. Rewrite the program that produces the animation associated with Fig. 11.6 but use separate animation instances for the movie on the left side and the histogram on the right.

4. Rewrite the program that produces the animation associated with

Animation

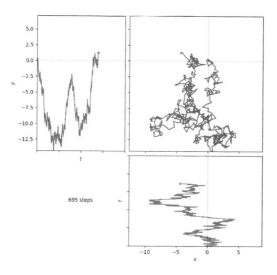

Fig. 11.7 but use a single animation instance for the movie on the left side and the animated plot on the right.

CHAPTER 12

Python Classes and GUIs

> *In this chapter, you learn how to create and use* **Python classes**, *which are central to what is known as* **object-oriented programming** *(OOP). You learn the basics of* **inheritance**, *a central concept in OOP, and how it can be employed to reuse code. You also learn the basics of how to write Python programs that the user can interact with through a* **graphical user interface**, *or GUI.*

Object-oriented programming—OOP—is an approach to programming centered around objects, which we introduced in §3.5. While using Python objects is relatively straightforward, object-oriented programming is a big subject and covering it thoroughly requires much more space than we can give it here. On the other hand, the basic machinery of OOP is relatively simple, especially in Python, and understanding it will give you a greater appreciation of how Python works.

An object, we learned, is a collection of data along with *methods* (functions), that can operate on that data, and *instance variables*, that characterize or are otherwise associated with the data. Taken together, these methods and instance variables are known as an object's *attributes*.

A NumPy array provides an illustrative example of an object. It contains data, in the form of the elements of the array, and it has a number of attributes, which can be accessed using the dot syntax. The size and shape of a NumPy array are examples of *instance variables* that are determined for a particular array when it is created, or *instantiated*, in the jargon of OOP:

```
In [1]: w = np.array([[2, -5, 6], [-10, 9, 7]])

In [2]: w.size      # size is an array instance variable
Out[2]: 6

In [3]: w.shape     # shape is another array instance variable
Out[3]: (2, 3)
```

NumPy arrays have methods associated with them, functions that act on a NumPy array, such as the methods that calculate the mean and standard deviation:

```
In [4]: w.mean()    # mean() is an array method
Out[4]: 1.5
```

```
In [5]: w.std()     # std() is another array method
Out[5]: 6.8495741960115053
```

Object methods always have parentheses, which may or may not take an argument. By contrast, instance variables do not have parentheses or take arguments.

In the language of OOP, we created an *instance* of the NumPy array *class* and named it w when we wrote w = np.array(...) above. Writing x = np.numpy([8, -4, -6, 3]) creates another instance of the NumPy array class, with different instance variables, but with the same set of methods (although using them on x would give different results than using them on w). w and x are two *objects* that belong to the same NumPy array class. Once we have instantiated an array, it is available for further queries or processing, which might involve interacting with other objects.

12.1 Defining and Using a Class

In Python, we can define new kinds of objects by writing classes to augment Python's classes, much like we can define our own functions to augment Python's functions.

To illustrate, we start by defining a class to model colloidal suspensions, which are very small microparticles, generally between a nanometer and a micrometer in diameter, suspended in a liquid. Colloidal gold was known to the Romans and used for staining glass. Today, colloids are valued for their medical uses and optical properties; they find application in a wide spectrum of technologies, including coatings and paints. Suspended in a liquid, colloidal microparticles are batted about by the liquid molecules, a phenomenon known as Brownian motion. This Brownian motion is characterized by a diffusion coefficient which can be determined using a formula Einstein derived in 1905: $D = k_B T/\zeta$, where k_B is Boltzmann's constant, T is the absolute temperature, and ζ is the friction coefficient. The friction coefficient is given by $\zeta = 3\pi\eta d$, where η is the viscosity of the suspending liquid, and d is the particle diameter. Under the influence of gravity, the particles tend to settle at a velocity of $v_{\text{sed}} = (\rho_p - \rho_l)g\pi d^3/6\zeta$, where g is the acceleration due to gravity, and ρ_p and ρ_l are the mass density of the particles and suspending liquid, respectively.

We define a `Colloid` class below, but before we describe the code for defining the class (on p. 318), let's see how it works. Our `Colloid` class takes five arguments: the particle diameter `pdiam` in meters, the particle density `pdens` in kg/m^3, the liquid viscosity `lvisc` in Pa-s, the liquid density `ldens` in kg/m^3, and the temperature `tempC` in degrees Celsius. All units are SI, meaning kilograms, meters, seconds, *etc.*[1] We design our `Colloid` class to have these as arguments to be input by the user:

```
Colloid(pdiam, pdens, lvisc, ldens, tempC)
```

As an example, let's define a gold (Au) colloid suspended in water (H$_2$O) at room temperature. Before getting started, however, we need to import the `Colloid` class into our IPython environment. The code for the `Colloid` class is stored in a file called `colloid.py`. This should be located in your IPython working directory. After putting it there, you can check to see if it's present by asking for a list of the files in your directory.

```
In [1]: ls
colloid.py
```

This tells us that the file `colloid.py` is present. Next, we need to import the `Colloid` class so that it's available to us to use:

```
In [2]: from colloid import Colloid
```

Note that we omit the `.py` extension on the file `colloid.py` that contains the class definition for the `Colloid` class. Now that we have imported the `Colloid` class, we can instantiate our gold colloid object.

```
In [3]: au_h2o = Colloid(7.5e-9, 19320., 0.00089, 1000., 25.)
```

Our `Colloid` class has six instance variables, which we access using the usual dot syntax, shown here in pairs to save space:

```
In [4]: au_h2o.pdiam, au_h2o.pdens    # diameter & density
Out[4]: (7.5e-09, 19320.0)            # of gold in m and kg/m^3

In [5]: au_h2o.lvisc, au_h2o.ldens    # diameter & density
```

---

[1] SI is abbreviated from the French *Système International* and simply refers to the modern metric systems of units.

```
Out[5]: (0.00089, 1000.0)          # of water in m and kg/m^3

In [6]: au_h2o.tempC, au_h2o.tempK # water temperature in
Out[6]: (25.0, 298.15)             # degrees Celsius & Kelvin
```

Note that our class has six instance variables but only five inputs. Obviously, the sixth, the temperature in Kelvin, is derived from the fifth, the temperature in degrees Celsius.

Our `Colloid` class also has several methods associated with it, which we illustrate here:

```
In [7]: au_h2o.pmass()             # particle mass in kg
Out[7]: 4.2676572703608835e-21

In [8]: au_h2o.vsed()              # particle sedimentation
Out[8]: 5.610499999999999e-10      # velocity in m/s

In [9]: au_h2o.diff_coef()         # particle diffusion
Out[9]: 6.303932584269662e-10      # coefficient in m^2/s
```

Like all classes, we can create many different instances of the class, each characterized by a different set of instance variables. For example, a class of polystyrene (plastic) colloids 0.5 μm in diameter with a density of 1050 kg/m^3 suspended in water can be instantiated:

```
In [10]: ps_h2o = Colloid(0.5e-6, 1050., 0.00089, 1000., 25.)
```

We can apply the same set of methods to this second `Colloid` object:

```
In [11]: ps_h2o.pmass()
Out[11]: 6.872233929727671e-17

In [12]: ps_h2o.vsed()
Out[12]: 7.64669163545568e-09
```

We have now created two instances—two objects—of the `Colloid` class: `au_h2o` and `ps_h2o`.

Now, let's examine the code that was used to define the `Colloid` class, which is given here.

Code: chapter12/programs/colloid.py

```
1  from numpy import pi, inf
2
3
4  class Colloid():
5      """A class to model a microparticle suspended in a liquid.
6      """
7
8      def __init__(self, pdiam, pdens, lvisc=0.00089,
```

Python Classes and GUIs 319

```
 9                ldens=1000., tempC=25.0):
10       """Initialize suspension properties in SI units."""
11       self.pdiam = pdiam   # particle diameter (m)
12       self.pdens = pdens   # particle density (kg/m^3)
13       self.lvisc = lvisc   # solvent viscosity (Pa-s)
14       self.ldens = ldens   # solvent density (kg/m^3)
15       self.tempC = tempC   # temperature (degrees C)
16       self.tempK = tempC + 273.15   # temperature (K)
17
18   def pmass(self):
19       """Calculate particle mass"""
20       return self.pdens*pi*self.pdiam**3/6.0
21
22   def friction(self):
23       return 3.0*pi*self.lvisc*self.pdiam
24
25   def vsed(self):
26       """Calculate particle sedimentation velocity"""
27       g = 9.80   # gravitational acceleration
28       grav = (pi/6.0)*(self.pdens-self.ldens)*g*self.pdiam**3
29       return grav/self.friction()
30
31   def diff_coef(self):
32       """Calculate particle diffusion coefficient"""
33       kB = 1.38064852e-23
34       return kB*self.tempK/self.friction()
```

The Colloid class is defined in line 4 by writing class Colloid() (the value of π and ∞ are imported from NumPy prior to the class definition because we need them later). The parentheses are empty because we are creating a new class, without reference to any pre-existing class. By convention, class names that we write are capitalized in Python, although it's not strictly required. We include a docstring briefly describing the new class.

12.1.1 The __init__() method

Looking over the Colloid class definition, we see a series of function definitions. These define the *methods* associated with the class. The __init__() method, which appears first, is a special method called the *constructor*. It has two leading and two trailing underscores to distinguish it from any other method you might define. The constructor is so named because it constructs (and initializes) an instance of the class. *This method is automatically called when you instantiate a class.*

The self argument must be specified as the first argument in the constructor, or `__init__` will not automatically be called upon instantiation. The constructor associates the name you give an instance of a class with the `self` variable, so you can think of it as the variable that identifies a particular instance of a class. So when we write `au_h2o = Colloid(...)`, the instance name `au_h2o` gets associated with the `self` variable, even though it doesn't appear as an explicit argument when you call it, as illustrated here:

```
au_h2o = Colloid(7.5e-9, 19320., 0.00089, 1000., 25.)
```

The five arguments above correspond to the five variables following the `self` argument in the `__init__()` definition. You undoubtedly noticed that the final three arguments of `__init__()` are provided with default values, as we might do for any function definition. Thus, because the default values correspond to those for water at room temperature, we could have instantiated `au_h2o` without specifying the last three arguments:

```
au_h2o = Colloid(7.5e-9, 19320.)
```

The body of the constructor initializes the class instance variables, which are all assigned names that have the `self` prefix. Any variable prefixed by `self` becomes an instance variable and is available to every other method in the class. The instance variables are passed to the other methods of the class through the argument `self`, which is the first argument of each method. The instance variables of one instance of a class are available only within that instance, and not to other instances of the same class. As noted previously, the instance variables can be accessed from the calling program using the dot syntax.

12.1.2 Defining methods for a class

Class methods are defined pretty much the same way any other function is defined. Class methods require `self` as the first argument, just as the `__init__` method does. This makes all instance variables of the class available to the method. When using the instance variables within a method, you must also use the `self` prefix.

Variables defined within a method that do not have the `self` prefix are local to that method and cannot be seen by other methods in the class or outside of the class. For example, Boltzmann's constant k_B, which is defined in the `diff_coef` method, is not available to the other methods because is does not have the `self` prefix. This is a good thing because it allows methods to have their own local variables (local namespaces).

The methods use the `return` statement to return values in exactly the same way conventional Python functions do.

12.1.3 Calling methods from within a class

Sometimes it may be convenient for one method in a class to call another method. Suppose, for example, we want to calculate the gravitational height of a colloidal suspension, which is given by the formula $h_g = D/v_{sed}$. Both D and v_{sed} are already calculated by the `diff_coef` and `vsed` methods, so we would like to write a method that simply calls these two methods and returns their quotient. Adding the following code to our class definition does the trick:

Code: chapter12/programs/colloid.py

```
36      def grav_height(self):
37          """Calculate gravitational height of particles"""
38          D = self.diff_coef()
39          v = self.vsed()
40          try:
41              hg = D/v
42          except ZeroDivisionError:
43              hg = inf    # when liquid & particle density equal
44          return hg
```

The methods `diff_coef` and `vsed` are called from within the class definition using the dot syntax exactly as they would be from without the class definition. The only difference is that the class instance name—say `au_h2o`, for example—is `self`, which makes sense since `self` serves as the instance name within the class definition. You may have also noticed that both the `diff_coef` and `vsed` methods called the `friction` method, again using the `self` prefix.

Note also that we created two local variables, `D` and `v`, and then returned their quotient. We define `D` and `v` without the `self` prefix because we wish to use them as purely local variables within the method `grav_height`. Defining `D` and `v` is done only for clarity to make it more evident to the reader what the calculation is doing.

A division by zero error can occur if $v_{sed} = (\rho_p - \rho_l)g\pi d^3/6\zeta = 0$, which occurs when the particle and liquid densities are equal ($\rho_p = \rho_l$). Lines 40–44 use Python's exception handling code, a `try`-`except` block, to deal with this possibility. The `try` block, starting at line 40, is attempted first. If a legal answer is obtained, the except block is skipped and the value of the *hg* calculated in the `try` block is returned. If there is a `ZeroDivisionError`, the except block is executed, *hg* is set to `inf` (infinity) and its result is returned. Other types of errors can also be handled with a `try-except` block. You can read about the various possibilities in Python's online documentation.

12.1.4 Updating instance variables

Instance variables are set when a class is instantiated. However, they are not immutable and can be changed after instantiation.

To demonstrate how this can be done, we first create a new instance of our `Colloid` class, this time for 0.62-μm-diameter particles made from a plastic-silica hybrid called TPC, which has a density of 1300 kg/m^3. The particles are suspended in liquid tetralin ($C_{10}H_{12}$), which has a density of 970 kg/m^3 and a viscosity of 0.202 Pa-s.

```
In [11]: tpc_C10H12 = Colloid(0.62e-6, 1300., 0.202, 970., 25.)
```

Let's check the particle diameter, then change it, and then check it again:

```
In [12]: tpc_C10H12.pdiam            # Check the diameter
Out[12]: 6.2e-07                     # Instantiated value

In [13]: tpc_C10H12.pdiam = 4.8e-7   # Set a new value

In [14]: tpc_C10H12.pdiam            # Recheck diameter
Out[14]: 4.8e-07                     # Set to new value
```

This shows that instance variables can be changed in a very straightforward manner.

Now let's change the temperature:

```
In [15]: tpc_C10H12.tempC = 35.

In [16]: tpc_C10H12.tempC
Out[16]: 35.0

In [17]: tpc_C10H12.tempK
Out[17]: 298.15
```

This worked as expected, but changed only the temperature in degrees Celsius; the temperature in Kelvin remained unchanged. To be consistent with the Celsius temperature, the Kelvin temperature should be 308.15, not 298.15. Obviously, this is a problem. To deal with it, we create a new method called `set_tempC` that can be used to change the temperature.

Code: chapter12/programs/colloid.py

```
46      def set_tempC(self, tempC):
47          """Sets temperature to a specfied value"""
48          self.tempC = tempC
49          self.tempK = tempC + 273.15
```

As an aside, we note that the new method `set_tempC` takes an argument, the temperature in degrees Celsius. The methods we previously wrote did not take any arguments (besides the internal argument `self`). Methods, like functions, can indeed take arguments, more than one in fact, in addition to `self`. As before, the `self` argument is not included in the method call. Now, let's try out the new method:

```
In [18]: tpc_C10H12.set_tempC(35.)

In [19]: tpc_C10H12.tempC
Out[19]: 35.0

In [20]: tpc_C10H12.tempK
Out[20]: 308.15
```

This gives the desired result. Methods like `set_tempC` that set one or more instance variables are called *setters* and can be very useful for ensuring the proper setting and updating of instance variables.

12.2 Inheritance

The `Colloid` class we made is pretty basic. The fact that it is basic has an upside and a downside. On the upside, because it is basic, it applies to virtually all colloids. On the downside, because it is basic, it misses properties of important types of colloids. It's for situations like this that the idea of class *inheritance* becomes useful. We can write a new class, called a *child* class, to take into account the unique properties of a certain kind of colloid, by building on the original *parent* class. We don't have to make the child class from scratch.

As an example, we are going to build a new child class called `HairyColloid` that's derived from the parent class `Colloid`. The colloids that the new class is meant to describe are not simply small hard spheres suspended in a liquid, they have short linear molecules tethered to their surfaces, "hairs" on the particles, that extend out into the liquid in which they are suspended. These tethered hairs inhibit aggregation of the particles, which is a good thing if the particles are to remain suspended. Hairy colloids are thus a specialized kind of colloid.

Besides keeping particles from aggregating, the main effect of the hairs tethered to the colloids is to slow how they move through the liquid in which they are suspended. If the length of the hairs is h, this change can be accounted for by introducing a *hydrodynamic diameter*, $d_h = d + 2h$, which is used in the calculation of the friction coefficient: $\zeta = 3\pi\eta d_h$. The diameter used to calculate the particle mass is unchanged and remains d.

To take this change into account in our new class `HairyColloid`, we will need to introduce two new instance variables, `hlen` and `hdiam`, which correspond to h and d_h, respectively, and we will need to redefine the `friction` method. Otherwise, we can reuse all the code from the original `Colloid` class definition.

Before coding our new `HairyColloid` class, we introduce some useful jargon: a parent class is often called the *superclass*; a child class is often called the *subclass*. Here, `Colloid` is the superclass; `HairyColloid` is the subclass.

Here is the code for the new subclass `HairyColloid`[2]:

Code: chapter12/programs/colloid.py

```
52  class HairyColloid(Colloid):
53      """A class to model hairy colloids"""
54
55      def __init__(self, pdiam, pdens, lvisc, ldens, tempC,
56                   hlen):
57          """Initialize properties from parent Colloid"""
58          super().__init__(pdiam, pdens, lvisc, ldens, tempC)
59          self.hlen = hlen    # length of hairs on particles
60          self.hdiam = pdiam + 2.0*hlen
61
62      def friction(self):
63          return 3.0*pi*self.lvisc*self.hdiam
```

You probably noticed that `Colloid` appears inside the parentheses for the class declaration of `HairyColloid` on line 52. This makes `HairyColloid` a subclass of `Colloid` and causes all the methods of `Colloid`, except `__init__()`, to become methods of `HairyColloid`. In our definition of `Colloid`, the parentheses in the class declaration were left empty, meaning that `Colloid` is not a subclass of any other class; it's an original.

---

[2] The code for `HairyColloid` follows the code for `Colloid` in the file `colloid.py`.

The `__init__()` declaration for `HairyColloid` includes all the arguments for the subclass, including those taken from the superclass. We don't have to explicitly reset all of the instance variables of the superclass one-by-one, however. We set the instance variables of the superclass using the method `super().__init__()`, where its arguments of `__init__()` are the arguments of the superclass (don't forget the two pairs of parentheses, one pair after `super` and the other after `__init__`). Finally, we initialize the new instance variables unique to `HairyColloid`: `self.hlen`, the length of the tethered hairs, and `self.hdiam`, the hydrodynamic diameter of a hairy colloid.

Next, we need to redefine the `friction` method so that it uses the hydrodynamic diameter. This is done in lines 62–63. This new `friction` method definition in `HairyColloid` replaces, or *overrides*, its definition in `Colloid`.

The other methods of `Colloid` remain unchanged and are available to `HairyColloid`. Below we try out our new subclass `HairyColloid` and compare it to its parent (or super) class `Colloid` by instantiating the two objects with the same inputs, except for the `hlen` argument (the last argument) unique to `HairyColloid`.

```
In [1]: au_h2o = Colloid(7.5e-9, 19320., 0.00089, 1000., 25.)

In [2]: au_h2o_hc = HairyColloid(7.5e-9, 19320., 0.00089,
   ...:                          1000., 25., 12.e-9)

In [3]: au_h2o.vsed(), au_h2o.diff_coef()
Out[3]: (6.303932584269662e-10, 6.543280646330727e-11)

In [4]: au_h2o_hc.vsed(), au_h2o_hc.diff_coef()
Out[4]: (1.5009363295880149e-10, 1.557923963412078e-11)

In [5]: au_h2o.grav_height()
Out[5]: 0.10379680554735492

In [6]: au_h2o_hc.grav_height()
Out[6]: 0.10379680554735493
```

The sedimentation velocities and diffusion coefficients for the two colloids are quite different, even though they have the same gravitational height. Note that Python returns values for the sedimentation velocity from the `Colloid` and `HairyColloid` classes that differ by about 1 part in 10^{16}. This minuscule difference comes from the very small roundoff error that occurs when the quotient `D/v` is calculated using the different values of `D` and `v` from the `au_h2o` and `au_h2o_hc` objects.

By the way, you can examine an object's instance variables with the __dict__ attribute:

```
In [7]: au_h2o_hc.__dict__
Out[7]:
{'hdiam': 3.15e-08,
 'hlen': 1.2e-08,
 'ldens': 1000.0,
 'lvisc': 0.00089,
 'pdens': 19320.0,
 'pdiam': 7.5e-09,
 'tempC': 25.0,
 'tempK': 298.15}
```

This brings us to the end of our very brief introduction to classes and OOP. Our next topic is graphical user interfaces or GUIs, which make extensive use of classes and is reason enough to learn the basics of OOP.

For a great deal of scientific programming, however, OOP is not needed and can even be an impediment sometimes, especially in Python. Making complex data structures with classes can make them difficult or impossible to vectorize using NumPy arrays, which is perhaps the most important tool available for scientific programming in Python, as it enables Python to perform scientific computations at speeds approaching those of compiled languages. So when speed is paramount, as it often is in scientific computing, you should use them judiciously in ways that do not seriously slow down computations.

12.3 Graphical User Interfaces (GUIs)

In this section, you learn how Python can be used create graphical user interfaces, or GUIs. GUIs can be enormously useful for interacting with software in a transparent and efficient manner. The subject, like OOP, is vast, so here we provide only the most basic of introductions to the subject. The hope is that this introduction will get you over the "hump" and give you the basic tools to launch out on your own and create GUIs for your own applications.

Our first project will be to make a GUI for colloids that provides the same information provided by the `HairyColloid` class we developed in the previous section, but with a more useful and user-friendly interface. The end result is shown in Fig. 12.1. There are six numerical inputs, starting with the particle diameter and ending with the

Python Classes and GUIs

brush length, and four numerical outputs, starting with the hydrodynamic diameter and ending with the gravitational height. The great strength of a GUI like this is that all the inputs and outputs are immediately accessible to the user. Moreover, the outputs immediately update when the value of input is changed.

The program we will write is meant to be run from the Terminal application, rather than an IDE like Spyder, or even from a Jupyter notebook which, of course, is more convenient for the user. We will show a work-around for running it from an IDE, but that is not how it is intended to be used.

12.3.1 Event-driven programming

A GUI program runs in a fundamentally different way than the programs we have encountered so far. Up until now, all of the programs we've encountered define some task, say plotting a figure, processing some data, or running a simulation, and then proceed through a series of steps until the task is finished. These programs may ask for some input from the user, but once it's obtained, the program simply runs through a set of calculations or procedures until it completes the task it was written to perform.

With a GUI program like the one introduced above, the program spends most of its time waiting for some new input from the user. Here, this occurs when the user changes a value in one of the input boxes, which are called *spinboxes*. The user can change the value in a spinbox from the keyboard by writing directly to the white area or by using the computer's mouse to click on the up or down arrows on the right side of the spinbox to increase or decrease its numerical

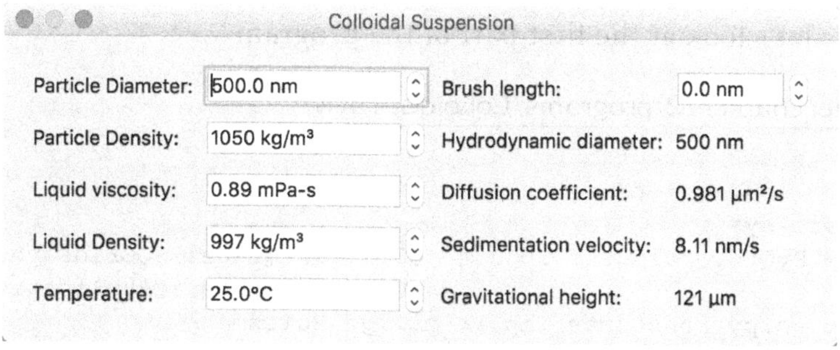

Figure 12.1 Colloid GUI window: inputs in white boxes, outputs on gray.

value. Either way, the program responds by redoing the calculations that determine the values of the outputs.

Programs like this are said to be *event-driven*. We have more to say about this later.

12.3.2 PyQt

Because a GUI interface is intrinsically graphical, we need software that can write directly to a computer screen. The software should work equally well, without modification, on Macs, PCs, and Linux computers. The software we choose to use for this task is called *PyQt*,[3] which is a Python wrapper for Qt, a C++ toolkit for writing graphics to computer screens on different platforms. We previously encountered Qt in Chapter 6 in our discussion of backends for matplotlib (see p. 133).

PyQt is a set of classes for performing all the tasks required for making a GUI. It makes simple dialog boxes, like the one shown in Fig. 12.1, as well as much more complicated ones. It provides tools that wait for events, such as a mouse click or input from a keyboard, as well as tools for processing (or acting upon) these events.

12.3.3 A basic PyQt dialog

We dive right in by looking at the code that produces the dialog shown in Fig. 12.1. The name of our program is `Colloid.pyw`. The full program is fairly long so we include just the first part to start our discussion. We import the `sys` module, which will be needed to launch the PyQt application. Next we import several PyQt5 classes which we will need for our program. Finally, we import some constants and functions from SciPy and NumPy that are needed for our calculations. Now let's look at the first part of the program.

Code: chapter12/programs/ColloidGUI.pyw

```
1  #!/usr/bin/env python3
2  # -*- coding: utf-8 -*-
3  import sys
4  from PyQt5.QtWidgets import (QDialog, QLabel, QGridLayout,
5                                QDoubleSpinBox, QApplication)
6  from scipy.constants import pi, g, Boltzmann
```

---

[3] PyQt usually comes pre-installed with the Anaconda distribution. Check that it is version 5, as the code we develop only works for PyQt5.

```python
 7  from numpy import abs, inf
 8
 9
10  class Form(QDialog):
11
12      def __init__(self):
13          super().__init__()
14          # inputs
15          self.pdiamSpinBox = QDoubleSpinBox()
16          self.pdiamSpinBox.setRange(1., 50000.)
17          self.pdiamSpinBox.setValue(500)
18          self.pdiamSpinBox.setSuffix(" nm")
19          self.pdiamSpinBox.setDecimals(1)
20          pdiamLabel = QLabel("Particle Diameter:")
21
22          self.pdensSpinBox = QDoubleSpinBox()
23          self.pdensSpinBox.setRange(1., 50000.)
24          self.pdensSpinBox.setValue(1050.)
25          self.pdensSpinBox.setSuffix(" kg/m\u00B3")
26          self.pdensSpinBox.setDecimals(0)
27          pdensLabel = QLabel("Particle Density:")
```

Making a dialog box

Making a dialog box starts by defining a class, which we name `Form`. `Form` inherits `QDialog` from the module `QtWidgets` of the `PyQt5` package. `QDialog` contains within it all the software needed to create our dialog box, which is done rather transparently, as we shall see below. The constructor for the `Form` class initializes the methods from the `QDialog` class with the `super().__init__()` call.

Making spinboxes and labeling them

The code for producing each spinbox is pretty much the same so we show here the code for only the first two spinboxes. The code for the others is pretty much the same.

The first block of code in lines 15–20 sets up the Particle Diameter spinbox. `QDoubleSpinBox()` creates a spinbox that takes floating point numbers, floats (or doubles in C++ syntax), as inputs. By contrast, `QSpinBox()`, not used here, takes integer inputs. The next four lines, 15–19, call various methods of `QDoubleSpinBox()` that, respectively, set the range of allowed inputs, sets the initial value, gives the input a suffix for the units for the input, and sets the number of decimal places that are displayed. In line 20, we call the `QLabel` routine, which defines a label for this spinbox that we will place next to the spinbox. The placement of the spinboxes and their labels is deferred until later in the program.

The second block of code in lines 22–27 sets up the second spinbox using the same set of calls used for the first spinbox. The superscript "3" in kg/m^3 is inserted using the Unicode escape sequence \u00B3. Tables of Unicode character codes can be found on the internet. You can also put in a Unicode superscript "^{3}" literal—that works too.

The remainder of the spinboxes are set up in the same way.

Outputs and their labels

The next set of code in lines 57–64 sets up the outputs, including their labels, that appear on the right side of the dialog. Line 57 defines the output variable `self.hydroDiam`, whose numerical value will be calculated later in the program. For the moment, we leave its value blank. Line 58 defines the variable `hydroDiamLabel` which we set to be the string `"Hydrodynamic diameter:"`. This will serve as the label for the value of the hydrodynamic diameter. Lines 59–64 define the outputs and labels for the other three outputs in the same way.

Code: chapter12/programs/ColloidGUI.pyw

```
57      self.hydroDiam = QLabel()
58      hydroDiamLabel = QLabel("Hydrodynamic diameter:")
59      self.diffCoef = QLabel()
60      diffCoefLabel = QLabel("Diffusion coefficient:")
61      self.vsed = QLabel()
62      vsedLabel = QLabel("Sedimentation velocity:")
63      self.gravHeight = QLabel()
64      gravHeightLabel = QLabel("Gravitational height:")
```

Laying out the inputs, outputs, and their labels

Lines 66–89 define the physical layout of the various elements of the window: the inputs, outputs, and their labels. The `PyQt5.QtWidgets` module contains a number of routines for laying out a window. Here, we use `QGridLayout`, which places the various elements of the window in a grid. We create an instance of `QGridLayout` and give it the variable name `grid`.

Code: chapter12/programs/ColloidGUI.pyw

```
66      grid = QGridLayout()
67      # inputs
```

```
68          grid.addWidget(pdiamLabel, 0, 0)        # Diameter
69          grid.addWidget(self.pdiamSpinBox, 0, 1)
70          grid.addWidget(pdensLabel, 1, 0)         # Density
71          grid.addWidget(self.pdensSpinBox, 1, 1)
72          grid.addWidget(lviscLabel, 2, 0)         # Viscosity
73          grid.addWidget(self.lviscSpinBox, 2, 1)
74          grid.addWidget(ldensLabel, 3, 0)         # Densiity
75          grid.addWidget(self.ldensSpinBox, 3, 1)
76          grid.addWidget(tempLabel, 4, 0)          # Temp (C)
77          grid.addWidget(self.tempCSpinBox, 4, 1)
78          grid.addWidget(brushLabel, 0, 2)         # Brush length
79          grid.addWidget(self.brushSpinBox, 0, 3)
80          # outputs
81          grid.addWidget(hydroDiamLabel, 1, 2)     # Hydro diam
82          grid.addWidget(self.hydroDiam, 1, 3)
83          grid.addWidget(diffCoefLabel, 2, 2)      # Diff coef
84          grid.addWidget(self.diffCoef, 2, 3)
85          grid.addWidget(vsedLabel, 3, 2)          # Sed vel
86          grid.addWidget(self.vsed, 3, 3)
87          grid.addWidget(gravHeightLabel, 4, 2)    # Grav height
88          grid.addWidget(self.gravHeight, 4, 3)
89          self.setLayout(grid)
```

We use the `addWidget` method of `QGridLayout` to place the various elements of the window. The `addWidget` method takes three arguments: the name of the element to be placed, the row number in the grid, and the column number. In this case we want to place objects in a grid of four columns and five rows, which are specified and placed as follows:

0, 0	0, 1	0, 2	0, 3
1, 0	1, 1	1, 2	1, 3
2, 0	2, 1	2, 2	2, 3
3, 0	3, 1	3, 2	3, 3
4, 0	4, 1	4, 2	4, 3

Note that in the `addWidget` calls, the input and output *variables* all have the `self` prefix, while their labels do not. That is because the output variables are set and updated in another method of our `Form` class. By contrast, their labels are set once when the class is initialized and do not change thereafter.

The `setLayout(grid)` call finishes the job of setting up the layout defined by `QGridLayout` and the calls to its `addWidget` method.

Setting up the event loop: Signals and slots

The code in lines 91–96 of the `Form` class constructor sets up the event loop. Line 91 specifies that if the value of the `pdiamSpinBox` spinbox is changed, then the `updateUi` method is called. Similarly, each of the lines 92–96 specifies that if a value of any of the other spinboxes is changed, then the `updateUi` method is called.

Code: chapter12/programs/ColloidGUI.pyw

```
91      self.pdiamSpinBox.valueChanged.connect(self.updateUi)
92      self.pdensSpinBox.valueChanged.connect(self.updateUi)
93      self.lviscSpinBox.valueChanged.connect(self.updateUi)
94      self.ldensSpinBox.valueChanged.connect(self.updateUi)
95      self.tempCSpinBox.valueChanged.connect(self.updateUi)
96      self.brushSpinBox.valueChanged.connect(self.updateUi)
97      # Window title & initialize values of outputs
98      self.setWindowTitle("Colloidal Suspension")
99      self.updateUi()
```

The syntax of these statements works like this. The `valueChanged` part of each statement specifies that when the value of a spinbox is changed, a *signal* is generated. The `connect(self.updateUi)` method specifies that when that particular signal is generated, it is processed by the `updateUi` method, which in the jargon of PyQt is called a *slot*. Each *signal* is thus connected to a particular *slot*. In this simple case, the same slot is used to process each of the signals, but in principle, different slots, *i.e.*, different methods, could be specified to process different signals.

Completing the constructor

The penultimate statement of the constructor, line 98, sets the title of the dialog box. The last statement of the constructor, line 99, calls the method `updateUi`, which calculates the initial values of the outputs based on the initial values of the inputs set in the `setValue` method calls in the first part of the constructor.

Processing the signals

In this example, a single routine `updateUi` processes all the signals. The routine calculates the four outputs from the six inputs. To calculate the outputs, it first converts all the inputs, passed to `updateUi` via the `self` prefix, to local variables in SI units.

Python Classes and GUIs

Code: chapter12/programs/ColloidGUI.pyw

```
101     def updateUi(self):
102         tempK = self.tempCSpinBox.value()+273.15
103         eta = self.lviscSpinBox.value()*0.001          # SI units
104         pdiam = self.pdiamSpinBox.value()*1e-9         # SI units
105         pdens = self.pdensSpinBox.value()              # SI units
106         ldens = self.ldensSpinBox.value()              # SI units
107         hdiam = pdiam+2.0e-9*self.brushSpinBox.value()     # SI
108         friction = 3.0*pi*eta*hdiam
109         D = Boltzmann*tempK/friction
110         vsed = (pi/6.0)*(pdens-ldens)*g*pdiam**3/friction
111         try:
112             hg = D/vsed      # gravitational height in SI units
113         except ZeroDivisionError:
114             hg = inf    # when liquid & particle density equal
115         self.diffCoef.setText("{0:0.3g} \u03BCm\u00B2/s"
116                               .format(D*1e12))
117         self.vsed.setText("{0:0.3g} nm/s".format(vsed*1e9))
118         self.hydroDiam.setText("{0:0.3g} nm"
119                                .format(hdiam*1e9))
120         # Set gravitational height, with exception for vsed=0
121         if abs(hg) < 0.001:   # small values in microns
122             self.gravHeight.setText("{0:0.3g} \u03BCm"
123                                     .format(hg*1e6))
124         elif abs(hg) < inf:   # large values in millimeters
125             self.gravHeight.setText("{0:0.3g} mm"
126                                     .format(hg*1e3))
127         else:                  # infinity (\u221E)
128             self.gravHeight.setText("\u221E")
129         return
```

The outputs are then calculated in SI units, but returned to their respective outputs in more convenient units that avoid large or small exponents. We include an exception handler, as we did previously in §12.1.3, to handle the divide-by-zero error that occurs when calculating the gravitational height for the case that the sedimentation velocity is zero.

The values of the outputs are set using the setText method of QLabel. The setText method takes a formatted string as input, so it's straightforward to include the appropriate units with each output. The gravitational height h_g is proportional to d^{-3}, where d is the particle diameter, which means that h_g can span a much larger range of numerical values as d changes than the other outputs, from fractions of a micron to hundreds of millimeters. To deal with this, we use an if statement that gives the gravitational height in millimeters if it is large and in microns if it is small.

With a `return` statement, the definition of the `Form` class is complete.

Launching the program and the event loop

The last four lines of code in `ColloidGUI.pyw` launch the program.

Code: chapter12/programs/ColloidGUI.pyw

```
132  app = QApplication(sys.argv)
133  form = Form()
134  form.show()
135  sys.exit(app.exec_())
```

The first of these lines, 132, creates an application object, which every PyQt5 application must do. The `sys.argv` argument accepts a list of arguments from the command line. In this case, we do not provide the application with any inputs from the command line. Nevertheless, the `sys.argv` argument is still required, even for zero inputs.

Line 133 creates an instance of the `Form` class and line 134 creates the dialog widget in memory ready to be displayed on screen.

The call `app.exec_()` in line 135 launches the application, which enters the event loop.[4] Embedding it as the arguments of `sys.exit()` ensures an orderly exit when the dialog box is closed.

When launched, a `ColloidGUI` dialog box appears, similar to the one shown in Fig. 12.1, but with the inputs set to their initial values and with the outputs calculated from those values. After that, the application waits for a change in one of the inputs, that is, in one of the spinboxes. Superficially, the event loop functions just like a while loop. Unlike a typical while loop, which occupies the computer processor (CPU) by checking the loop variable for a change, the program doesn't occupy the CPU at all unless an event occurs. In this case, when you use the keyboard or mouse to change the value of a spinbox, the hardware generates an *interrupt*, causing the CPU to suspend what it is doing and process the interrupt, which in this case means reading the new value in the spinbox, generating a signal for our event loop, which then processes it according to the slot assigned to that signal. Here, all the signals go to the same slot, `updateUi`. Once the event is processed, our application once again quietly waits for another event, while the CPU returns to its other business.

---

[4]The underline is included in `exec_()` to avoid conflicts with the built-in Python function `exec()`.

Python Classes and GUIs

Running the program from the terminal

The `ColloidGUI.pyw` program, as written, is meant to run from the terminal and not from an IDE like Spyder. The next section will introduce a modification that will allow it to run from an IDE.

A second, and better, way to run the program is to make the application an executable program.

Making a program executable on a Mac

On Macs and on Linux computers, you can make `ColloidGUI.pyw` executable by typing from the terminal

```
chmod +x ColloidGUI.pyw
```

The next step is to include as the first line of your program

```
#!/usr/bin/env python3
```

This line tells the operating system to run this program with the `python3` directive located in the directory `/usr/bin/env`. Thus, while Python ignores this statement because it starts with the comment character `#`, the operating system reads it as an instruction of how to execute the program. The combination `#!` that alerts the system that a system directive follows is called a *shebang* statement.

Including the shebang statement and having made the program executable, simply typing `./ColloidGUI.pyw` from the terminal will run the program, if the program is in the same directory. The program can be run from any directory provided `ColloidGUI.pyw` is located in a directory that is in the system's `PATH` variable. This can be done by including the following line the file `.bash_profile`

```
export PATH="/Users/pine/scripts:$PATH"
```

which makes any program executable if it is located in the directory `/Users/pine/scripts`. In this case, that means that typing `ColloidGUI.pyw` from the terminal will run the program, irrespective of the current directory.

The file `.bash_profile` is located in your computer's default directory, which in the above example is `/Users/pine`. However, editing `.bash_profile` can be tricky as files with names that begin with a period are invisible in the Mac Finder window and also to many text editors. To edit the `.bash_profile`, open the Terminal application and navigate to the home directory, which is `/Users/pine` on my Mac; yours will be different. Then type `open .bash_profile`. This will open the file `.bash_profile` with Mac's default TextEdit program. Add the necessary line and save the file. Alternatively, you can use some other text editor that allows you to edit files that begin with a period.

One note of caution: hidden files like `.bash_profile` are hidden for a reason. They contain system configuration information that, if incorrectly changed, can cause applications not to function properly. Therefore, be careful and make backups before changing the `.bash_profile` file.

Finally, it's worth noting that you can run multiple instances of the `ColloidGUI.pyw`. On a Mac, simply type

```
ColloidGUI.pyw &
```

This launches `ColloidGUI.pyw` and returns the terminal to its usual $ prompt, ready to accept further system commands. Thus, typing `ColloidGUI.pyw &` again launches another instance of the GUI application. This is a general feature of the Unix operating system used by Macs. The same syntax works on computers running Linux as well.

Running the program from an IDE

Common IDEs like Spyder, as well as Jupyter notebooks, run an IPython shell. The PyQt `QApplication` will usually run once in an IPython shell, but if you run it a second time you get an error message

```
QCoreApplication::exec: The event loop is already running
```

then, most likely, Python crashes. The problem is that an instance of `QApplication` remains in the namespace of IPython after the first run. In the code below, we remedy this by checking to see if a `QApplication` instance already exists with the method `instance()`.

```python
if not QApplication.instance():
    app = QApplication(sys.argv)
else:
    app = QApplication.instance()
form = Form()
form.show()
sys.exit(app.exec_())
```

Running a GUI application from an IDE is a bit of an odd thing to do, as the program is meant to be run on its own, so whether or not you choose to use code like that above is your choice. Running it from a terminal is the standard protocol.

12.3.4 Summary of PyQt5 classes used

Here we summarize the PyQt5 classes and methods we have introduced, plus a few more. The ones we mention are but a small fraction of the classes defined in PyQt5, as we have given only the briefest of introductions to PyQt5. Nevertheless, to organize your thoughts, it is useful at this point to list a few of the PyQt5 classes and their methods.

Method	Function
`QLineEdit()`	Single-line textbox for input
`QTextBrowser()`	Multiline textbox for output (plain & HTML)
`QVBoxLayout()`	Lays out widgets vertically in a box
`QHBoxLayout()`	Lays out widgets horizontally in a box
`addWidget()`	method to add widgets to `Q..BoxLayout()`

Table 12.1   A selection of QtWidgets methods.

12.3.5 GUI summary

We have barely scratched the surface of the kinds of GUIs that can be made using Python. We have not, for example, said anything about how matplotlib plots can be incorporated into GUIs, nor have we shown you how to interact with plots and GUIs using a mouse. All of this and much more is possible. Our purpose here has been to introduce classes and how they are used to make GUIs. Armed with ideas, you are now equipped to read online documentation and take the next steps to create applications that suit your purposes.

APPENDIX A

Installing Python

For scientific programming with Python, you need to install Python and four scientific Python libraries: NumPy, SciPy, matplotlib, and Pandas. There are many other useful libraries you can install, but these four are probably the most widely used and are the only ones you will need for this text.

In this text, we work exclusively with Python 3.x and not with Python 2.7 or earlier versions of Python. When you download Python, be sure you **download Python 3.x** (x should be 6 or greater).

A.1 Installing Python

There are a number of ways to install Python and the scientific libraries you will need on your computer. Some are easier than others.

For most people, the simplest way to install Python and all the scientific libraries you need is to use the *Anaconda* distribution, which includes the Spyder integrated development environment (IDE) for Python. The **Anaconda distribution** also includes Jupyter notebooks for those who prefer to interact with Python using a web-based notebook format (see Appendix B for an introduction to Jupyter notebooks). Both of these interfaces feature syntax highlighting, which colors different parts Python syntax according to function, making code easier to read. More importantly, the Spyder IDE runs a program in the background called *Pyflakes* that checks the validity of the Python syntax as you write it. It's like a spelling and grammar checker all rolled into one, and it is extremely useful, for novice and expert alike. The Spyder IDE has a number of other useful features, which we do not go into here, but expect you will learn about as you become more familiar with Python.

Spyder provides an open source programming environment for Python. The Spyder IDE is free and is bundled with the Anaconda distribution of Python in a package maintained by the software company *Continuum Analytics* (https://www.anaconda.com/). The Anaconda package can be found at https://www.anaconda.com/download/. Be sure to **download Python 3.x** (x should be 6 or greater) and not Python 2.7.

In this book, we assume you are using either the Spyder IDE or a Jupyter notebook. Both are similar enough that we generally do not specify one or the other after Chapter 2. Once you install the Anaconda distribution, you can leave Spyder with its default settings, but we suggest a few changes to the default settings, which are detailed below.

A.1.1 Setting preferences

In this manual we assume that your **IPython Console** is set up with "pylab mode" turned off. This means that in the IPython Console you need to use the `np` or `plt` prefixes with NumPy and matplotlib functions.

We recommend that you use the Qt5Agg backend, although it's not strictly necessary for most of what we do.

Note that you only need to do this setup once. Once it is done, you should be able to follow everything written in this manual.

Launch Spyder and then go to the Preferences menu (under the python menu on a Mac, or the Tools on a PC). Under the *IPython Console: Graphics* menu, change the *Backend* to *Qt5* (it may already be selected, in which case you need to do nothing). Next, in the *IPython Console: Graphics* menu, make sure the box labeled Automatically load PyLab and NumPy modules is not checked. This sets up Spyder so that NumPy or matplotlib are not automatically loaded in the IPython console.

As of this writing, there is a bug in the Mac interface that omits the × on the tabs for different files in the Code Editor Pane (see Fig. 2.1. Clicking on the × (when it was working) closes the tab. There are two work-arounds to deal with this bug. (1) Just ignore it and understand that clicking on the left side of the tab will close the tab. (2) Go to the Preferences menu of Spyder and select the General tab. Then under Interface: Qt windows style, change "Macintosh" to "Fusion" or "Windows".

A.1.2 Pyflakes

A syntax-checking program called *Pyflakes* runs in the background when you are editing a Python program using Spyder. If there is an

Installing Python

error in your code, Pyflakes flags the error. See §2.11.1 for more information about Pyflakes and how to take advantage of its features.

A.1.3 Updating your Python installation

Updating the Anaconda distribution of Python, including Spyder, is straightforward using a terminal window. On a Mac, go to the Application folder and launch the Terminal app located in the Utilities folder. On a PC, go to the Start menu and launch the Anaconda Prompt app under the Anaconda (or Anaconda3) menu. From the terminal prompt, type

```
conda update conda
conda update anaconda
```

Respond [y]es to any prompts you receive to download and update your software. Do this once per month or so to keep Python and all the Python packages you have loaded up to date.

A.2 Testing Your Installation of Python

Running the program testInsatllation.py below tests your installation of Python and records information about the installed versions of various packages that are used in this manual. If you are a student, you should input your first and last names inside the single quotes on lines 10 and 11, respectively. Instructors should modify the course information inside the double quotes in lines 15-17.

Code: Appendix/programs/testIstallation.py

```
1  """ Checks Python installation and generates a pdf image file
2      that reports the versions of Python and selected installed
3      packages.  Students can sent output file to instructor."""
4  import scipy, numpy, matplotlib, pandas, platform, socket, sys
5  import matplotlib.pyplot as plt
6
7  # If you are a student, please fill in your first and last
8  # names inside the single quotes in the two lines below.
9  # You do not need to modify anything else in this file.
10 student_first_name = 'Giselle'
11 student_last_name = 'Sparks'
12 # If you are an instructor, modify the text between the
13 # double quotes on the next 3 lines. You do not need to
14 # modify anything else in this file.
15 classname = "Quantum Mechanics I"
```

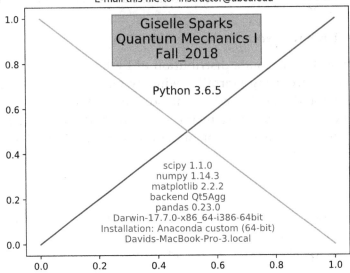

```
16   term = "Fall_2018"    # must contain no spaces
17   email = "instructor@abcu.edu"
18   plt.plot([0, 1], "C0", [1, 0], 'C1')
19   plt.text(0.5, 1.0, "{0:s} {1:s}\n{2:s}\n{3:s}"
20            .format(student_first_name, student_last_name,
21                    classname, term),
22            ha="center", va="top", size='x-large',
23            bbox=dict(facecolor="C2", alpha=0.4))
24   plt.text(0.5, 0.7,
25            'Python {0:s}'.format(platform.python_version()),
26            ha='center', va='top', size='large')
27   pkgstr = 'scipy {0:s}\nnumpy {1:s}\nmatplotlib {2:s}\n'
28   pkgstr += 'backend {3:s}\npandas {4:s}\n{5:s}\n'
29   pkgstr += 'Installation: {6:s}\n{7:s}'
30   plt.text(0.5, 0.0, pkgstr.format(scipy.__version__,
31            numpy.__version__, matplotlib.__version__,
32            matplotlib.get_backend(), pandas.__version__,
33            platform.platform(), sys.version.split('|')[1],
34            socket.gethostname()), ha='center', va='bottom',
35            color='C5')
36   filename = student_last_name+'_'+student_first_name
37   filename += '_'+term+'.pdf'
38   ttlstr = 'This plot has been saved on your computer as'
39   ttlstr += '\n"{0:s}"\nE-mail this file to "{1:s}"'
40   plt.title(ttlstr.format(filename, email), fontsize=10)
```

```
41  plt.savefig(filename)
42  plt.show()
```

A.3 Installing FFmpeg for Saving Animations

To record animations to an independent movie file, you need to install some external software. We suggest using FFmpeg, which works nicely with matplotlib.

Installing FFmpeg is a fairly simple matter. Go to your computer's terminal application (Terminal on a Mac or Anaconda Prompt on a PC) and type:

```
conda install -c anaconda ffmpeg
```

Respond [y]es when asked to proceed. The Anaconda utility `conda` will install a number of new packages and will probably update others. When it's finished, FFmpeg should work with the matplotlib animation programs discussed in Chapter 11.

APPENDIX B

Jupyter Notebooks

A Jupyter notebook is a web-browser-based environment for interactive computing. You don't need to be connected to the web to use it; Jupyter merely runs on your browser. If you have installed the Anaconda distribution and you have a standard web browser, then you have everything you need to launch a Jupyter notebook.

You can work in a Jupyter notebook interactively, just as you would using the IPython shell. In addition, you can store and run programs in a Jupyter notebook just like you would within the *Spyder* IDE. Thus, it would seem that a Jupyter notebook and the *Spyder* IDE do essentially the same thing. Up to a point, that is true. *Spyder* is generally more useful for developing, storing, and running code. A Jupyter notebook, on the other hand, is excellent for logging your work in Python. For example, Jupyter notebooks are very useful in a laboratory setting for reading, logging, and analyzing data. They are also useful for logging and turning in homework assignments. You may find them useful in other contexts for documenting and demonstrating software.

There is important advantage of using the Spyder IDE compared to using a Jupyter Notebook, especially for a newcomer to Python. Spyder runs a syntax checking program called Pyflakes in the background as you write and flags programming errors. Pyflakes can be implemented in Jupyter but using it is not automatic and a bit clumsy. Therefore, you are advised to start learning Python with Spyder before taking advantage of Jupyter.

B.1 Launching a Jupyter Notebook

To launch a Jupyter notebook, launch the Terminal (Mac) or the Anaconda Command Prompt (PC) application. On a Mac, the Terminal application is found in the Applications/Utilities folder. On a PC, the Anaconda Command Prompt application is found in the Start/All Programs/Accessories menu. Here we will refer the Terminal or Anaconda Command Prompt applications at the System Console. Once

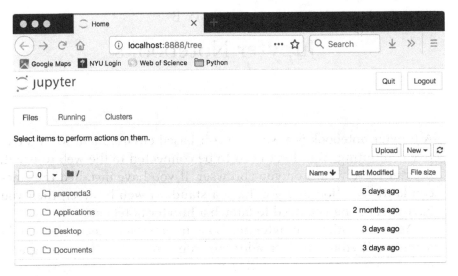

Figure B.1 Jupyter notebook dashboard.

you are in the System Console, type `jupyter notebook`. This will launch the Jupyter notebook web application and will display the *Jupyter Notebook Dashboard* as a page in your default web browser. It should look like the web page shown in Fig. B.1.

At the bottom-left of the Notebook Dashboard is a list of folders (and files) of the default directory. By clicking on the appropriate folder you can navigate to the folder where you want to store your work.

If you want to create a new folder, you can do this by clicking on the New button near the top-right of the Notebook Dashboard and selecting Folder. This will create a new folder called Untitled Folder. Let's rename this folder. To rename it, click on the checkbox to the left of the Untitled Folder listing. Then click on the box Rename (which appears only after you check the checkbox) near the top left of the window. A dialog box will appear where you should enter the new name of the folder. Let's call it Notebooks. Next, double click on the Notebooks listing, which will move the Jupyter interface to that folder.

To create a new Jupyter notebook, go to the pull-down menu New on the right side of the page and select Python 3. That opens a new *Jupyter notebook* with the provisional title `Untitled0` in a new tab like the one shown in Fig. B.2. To give the notebook a more meaningful

Jupyter Notebooks

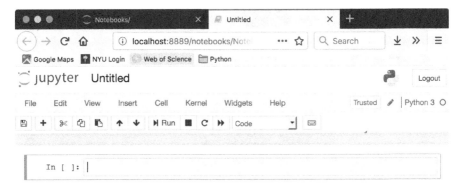

Figure B.2 Untitled Jupyter notebook with an open cell.

name, click on the File menu *in the browser window* and select Rename. Let's say you rename your notebook firstNotebook. The name firstNotebook will replace Untitled0 in your Notebook browser window and a file named FirstNotebook.ipynb will appear in the directory from which you launched Jupyter notebook. That file will contain all the work you do in the Jupyter notebook. Next time you launch Jupyter notebook from this same directory, all the Jupyter notebooks in that directory will appear in a list on the Jupyter notebook Dashboard. Clicking on one of them will launch that notebook.

When you open a new Jupyter notebook, an IPython interactive cell appears with the prompt In[ ]: to the left. You can type code into this cell just as you would in the IPython shell of the *Spyder* IDE. For example, typing 2+3 into the cell and pressing Shift-Enter (or Shift-Return) executes the cell and yields the expected result. Try it out.

If you want to delete a cell, you can do so by clicking on the cell and then selecting Delete Cells from the Edit menu. Go ahead and delete the cell you just entered. You can also restart a notebook by selecting Restart & Clear Output from the Kernel menu. Go ahead and do this too.

B.2 Running Programs in a Jupyter Notebook

You can run programs in a Jupyter notebook. As an example, we run the program introduced in §4.2.1. The program is input into a single notebook cell, as shown in Fig. B.3, and then executed by pressing Shift-Enter.

348   Introduction to Python for Science & Engineering

Figure B.3   Running a program in a Jupyter notebook.

The program runs up to the point where it needs input from the user, and then pauses until the user responds by filling in a distance and then pressing the Enter or Return key. The program then completes its execution. Thus, the Jupyter notebook provides a complete log of the session, which is shown in Fig. B.3.

B.3   Annotating a Jupyter Notebook

A Jupyter notebook will be easier to understand if it includes annotations that explain what is going on in the notebook. In addition to

Jupyter Notebooks

logging the inputs and outputs of computations, Jupyter notebooks allow the user to embed headings, explanatory notes, mathematics, and images.

B.3.1 Adding headings and text

Suppose, for example, that we want to have a title at the top of the Jupyter notebook we have been working with, and we want to include the name of the author of the session. To do this, we scroll the Jupyter notebook back up to the top and place the cursor in the very first input cell and click the mouse. We then open the Insert menu near the top center of the window and click on Insert Cell Above, which opens up a new input cell above the first cell. Next, we click on the box in the Toolbar that says Code. A list of cell types appears: Code (currently checked), Markdown, Raw NBConvert, and Heading. Select Markdown; immediately the In [ ]: prompt disappears, indicating that this box is no longer meant for inputting and executing Python code. Type "# Demo of Jupyter notebook" and press Return or Enter. Then type "## Your Name. Finally, press Shift-Enter (or Shift-Return). A heading in large print appears before the first IPython code cell, with "Your Name" printed below it in slightly smaller print because you typed ## instead of #. Each additional # decreases the font size.

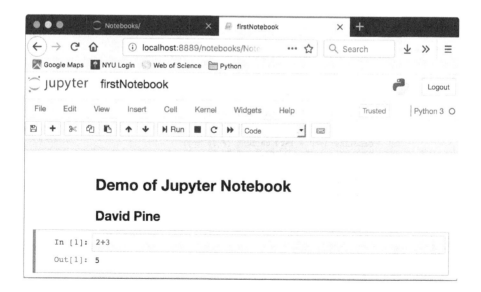

Figure B.4   Jupyter notebook with title and author.

B.3.2 Comments with mathematical expressions

You can also write comments, including mathematical expressions, in a Jupyter notebook cell. Let's include a comment after the program that calculated the cost of gasoline for a road trip. First we place the cursor in the open formula cell below the program we ran and then click on the box in the Toolbar that says Code and change it to Markdown. Returning to the cell, we enter the text of our comment. We can enter any text we wish, including mathematical expressions using the markup language LaTeX. (If you do not already know LaTeX, you can get a brief introduction at these sites: http://en.wikibooks.org/wiki/LaTeX/Mathematics or ftp://ftp.ams.org/pub/tex/doc/amsmath/short-math-guide.pdf.) Here we enter the following text:

```
The total distance $x$ traveled during a trip can be
obtained by integrating the velocity $v(t)$ over the
duration $T$ of the trip:
\begin{align}
```

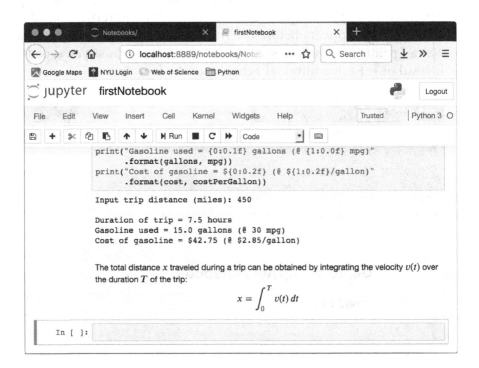

Figure B.5 Annotation using a Markdown cell with a mathematical expression.

```
            x = \int_0^T v(t)\, dt
\end{align}
```

After entering the text, pressing Shift-Enter yields the result shown in Fig. B.5.

The $ symbol brackets inline mathematical expressions in Latex, while the `\begin{align}` and `\end{align}` expressions bracket displayed expressions. You only need to use L^AT_EX if you want to have fancy mathematical expressions in your notes. Otherwise, LaTeX is not necessary.

B.4 Terminal commands in a Jupyter notebook

You can execute system shell commands from an IPython prompt, as we learned in n §2.4.2. Suppose, for example, you were importing a data (.txt) file from your hard disk and you wanted to print it out in one of the notebook cells. If you were in the Terminal (Mac) or Command Prompt (PC), you could write the contents of any text file using the command `cat` *filename* (Mac) or `type` *filename* (PC). You can execute the same operation from the IPython prompt using the Unix (Mac) or DOS (PC) command preceded by an exclamation point, as described in the section on *System shell commands*.

B.5 Plotting in a Jupyter Notebook

You can also incorporate matplotlib plots into a Jupyter notebook. At the `In [ ]:` prompt, type the following three lines:

```
In [3]:  import numpy as np
         import matplotlib.pyplot as plt
         %matplotlib inline
```

Then press `Shift-Enter` (or `Shift-Return`) to execute the code in the cell. The `%matplotlib inline` magic command tells the IPython to render the plots within IPython, that is in the Jupyter notebook, rather than in a separate window.[1]

---

[1] You can also use the `%matplotlib inline` magic command in the Spyder IPython window to get the same effect: displaying plots in the IPython shell instead of in a separate window.

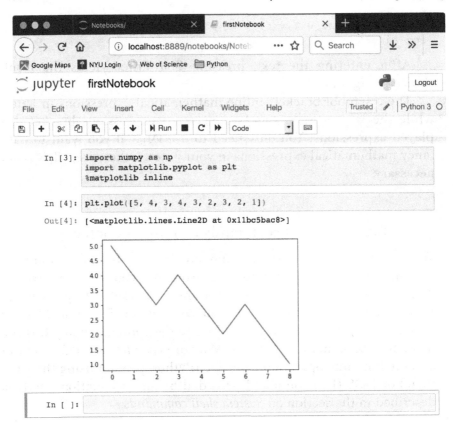

Figure B.6 Plot in a Jupyter notebook.

In the next cell, type `plt.plot([5, 4, 3, 4, 3, 2, 3, 2, 1])`. A plot should appear in the browser window beneath the `plot` command, like the one in Fig. B.6.

Let's review what just happened. In the first Jupyter notebook cell, we imported NumPy and PyPlot in the first two lines and gave them their standard prefixes `np` and `plt`. The third line is an IPython magic command (see §2.4.1) that instructs matplotlib to render plots "inline" in the Jupyter notebook instead of in a separate window.

In cell 2, we type a `plot` function, which is then output below in the notebook. Within the notebook with inline plotting, the `show()` function is not needed.

Be sure to press the `Save and Checkpoint` item in the `File` menu in the Jupyter notebook window from time to time to **save your work**.

B.6 Editing and Rerunning a Notebook

In working with a Jupyter notebook, you may find that you want to move some cells around, or delete some cells, or simply change some cells. All of these tasks are possible. You can cut and paste cells, as in a normal document editor, using the Edit menu. You can also freely edit cells and re-execute them by pressing Shift-Enter. Sometimes you may find that you would like to re-execute the entire notebook afresh. You can do this by going to the Kernel menu and selecting Restart. A warning message will appear asking you if you really want to restart. Answer in the affirmative. Then open the Cell menu and select Run All, which will re-execute the notebook starting with the first cell. You will have to re-enter any screen input requested by the notebook scripts.

B.7 Quitting a Jupyter Notebook

It goes almost without saying that before quitting a Jupyter notebook, you should make sure you have saved the notebook by pressing the Save and Checkpoint item in the File menu or its icon in the Toolbar.

When you are ready to quit working with a notebook, click on the Close and halt item in the File menu. Your notebook browser tab will close and you will return to the Jupyter notebook Dashboard. Press the Quit button at the upper right. Then close the IPython Notebook Dashboard tab in your browser to end the session.

Finally, return to the Terminal or Anaconda Command Prompt application. You should see the normal system prompt. If you don't, hold down the control key and press c twice in rapid succession. You can then close the Terminal (Mac) or Anaconda Command Prompt (PC) session if you wish.

B.8 Working with an Existing Jupyter Notebook

To work with an existing Jupyter notebook, open the Terminal (Mac) or Anaconda Command Prompt (PC) application and navigate to the directory in which the notebook you want to work with resides. Recall that Jupyter notebooks have the .ipynb extension. Launch the Jupyter notebook Dashboard as you did previously by issuing the command

```
jupyter notebook
```

This will open the Jupyter notebook Dashboard in your web browser, where you should see a list of all the Jupyter notebooks in that directory (folder). Click on the name of the notebook you want to open. It will appear in a new tab on your web browser as before.

Note that while all the input and output from the previous saved session is present, none of it has been run. That means that none of the variables or other objects has been defined in this new session. To initialize all the objects in the file, you must rerun the file. To rerun the file, press the Cell menu and select Run All, which will re-execute all the cells. You will have to re-enter any screen input requested by the notebook scripts. Now you are ready to pick up where you left off the last time.

APPENDIX C

Glossary

A number of terms that are introduced in the text are defined below for your convenience. The page number where the term is first used or defined is given in parentheses.

Artist (135) Artists in matplotlib are the routines that define the objects that are drawn in a plot: lines, circles, rectangles, axes, data points, legends, *etc*. The artist layer consists of the hierarchy of Python objects (or classes) that facilitate creating a figure and embellishing it with any and all of the features listed above.

Attributes (57) The methods and instance variables of an object.

Backend (132) A matplotlib backend translates plotting code into useful output. There are two types: hardcopy backends and interactive (alternatively called interface) backends. A hardcopy backend translates matplotlib code into image files such as PDF, PNG, PS, or SVG. An interactive backend translates matplotlib code into instructions your computer screen can understand. It does this using third-party cross-platform software, usually written in C or C++, that can produce instructions that are sent to your computer screen. The net result is matplotlib code that is platform-independent, working equally well under Windows, macOS, and Linux.

Blitting (298) Blitting most generally refers to the transfer of a block of pixel data from memory to your computer screen. However, in the context of animations, it refers to updating only those regions of the screen that change from one animation frame to the next. In an animated plot, such as the one displayed in Fig. 11.3 for example, blitting means only updating the plot itself since it is the only thing that is changing. The axes' labels and tick marks are not redrawn since they are not changing. Not redrawing static features like the axes, axes labels, and ticks can dramatically speed up an animation.

Dynamically typed language (1) In a *statically typed language*, variable names are declared to be of a certain type—say a floating point number *or* an integer—before they are ever used. A variable's type is fixed at the beginning of a program and cannot be changed during the execution of the program. In a *dynamically typed language*, variable names are not generally declared to be of a certain type. Instead their type is determined on the fly as the program runs. Moreover, the type can change during the execution of the program.

Instance variables (57) Data stored with an object that can be accessed by appending a period (.) followed by the name of the instance variable (without parentheses).

Instantiate (315) Create a new *instance* (realization) of a class by calling the class and providing the arguments required by the class constructor (__init__ method).

Method (55) Function associated with an object that acts on the object. A method is invoked by appending a period (.) followed by the name of the method and an open-close set of parentheses, which can, but need not, have an argument.

Object (2) In object-oriented programming, an object is generally thought of as a collection of data together with methods that act on the data and instance variables that characterize various aspects of the data.

Object-oriented programming (OOP) (2) Object-oriented programming refers to programming based on the concept of objects that interact with each other in a modular manner. Scientific programming is typically procedural. As the size and complexity of a scientific programming task increases, however, object-oriented design becomes increasingly useful.

Universal function or **ufunc** (161) A function that operates on NumPy arrays (ndarrays) in an element-by-element fashion. Such a function is said to be "vectorized." A NumPy ufunc also respects certain other rules about handling arrays with different sizes and shapes.

Vectorized code (133) Computer code that processes *vectors* (ndarrays in NumPy) as the basic unit rather than individual data elements.

Widget (161) In graphical-user interfaces (GUIs), a widget is a routine to create and define the function of a GUI element, such as a button, a spinbox, *etc.*

Wrapper (19) A Python program that provides a Python interface to a program written in another language, usually C, C++, or Fortran.

APPENDIX D

Python Resources

This text provides an introduction to Python for science and engineering applications but is hardly exhaustive. There are many other resources that you will want to tap. Here I point out several that you may find useful.

D.1 Python Programs and Data Files Introduced in This Text

Throughout the text, various Python programs and data files are introduced. All are freely available at https://github.com/djpine/python-scieng-public.

D.2 Web Resources

The best web resource for Python is a good search engine like Google. Nevertheless, I list a few web sites here that you might find useful.

Python home page: https://www.python.org/.
The official Python web site. I almost never look here.

Python 3 language reference: https://docs.python.org/3/reference/.
I look here sometimes for detailed information about Python 3, which is the version used in this text.

NumPy Reference: https://docs.scipy.org/doc/numpy/reference/.
I usually start here when I need information about NumPy. It has links to just about all the NumPy documentation I need. By the way, I say "num-pee," which rhymes with "bumpy"—a lot of people say "num-pie," which doesn't sound like English to me.

SciPy Reference: https://docs.scipy.org/doc/scipy/reference/.
I start here when I need information about SciPy, its various packages and their functions. I say "psy-pi" for SciPy, like everyone else. Who says I have to be consistent? (See Emerson.)

Pyplot: https://matplotlib.org/api/pyplot_summary.html.
The *Plotting Commands Summary* page for matplotlib. It has a search feature and links to all the matplotlib documentation, which I use a lot. You can go the main matplotlib page, http://matplotlib.org/, but frankly, it's less useful.

Stack Overflow: https://stackoverflow.com/questions/.
stack**overflow** is a set of web sites that lets people pose and answer questions related to computer programming in every computer language imaginable. Nearly every question or problem you can think of has already been asked and answered. Pose your question as specifically as you can, and find the solution to your problem. Take advantage of this valuable resource.

Pandas: http://pandas.pydata.org/pandas-docs/stable/10min.html.
There is probably some useful information here, but frankly, I recommend stack**overflow** for nearly all your questions about Pandas: https://stackoverflow.com/tags/pandas/info.

Jupyter notebooks: http://jupyter.org/.
I go to this page to learn about Jupyter notebooks. The site also has links to the IPython documentation.

Anaconda/Continuum: https://www.anaconda.com/.
Get the latest version of Spyder, and all the Python packages you want here, from Continuum Analytics. The Python package they maintain, which includes the Spyder IDE, is called Anaconda. Download the Anaconda distribution here: https://www.continuum.io/downloads/. Anaconda is completely open source and has a number of nice features, including Spyder which integrates documentation for the various Python packages with a simple cmd-I (Mac) or ctrl-I (PC & Linux) when the cursor is on the Python function under question. Anaconda's package manager is easy to use. See §A.1.3 for instructions about how to update Anaconda. Anaconda with Spyder is available on all platforms: Macs, PCs, and Linux machines.

Mailing lists: Some software packages have mailing lists to which you can subscribe or pose questions about a specific package. They give you access to a community of developers and users that can often provide expert help. Just remember to be polite and respectful of those helping you and also to those posting questions. The URL for the SciPy mailing list is http://www.scipy.org/Mailing_Lists/. The URL for the matplotlib mailing list is https://lists.sourceforge.net/lists/listinfo/matplotlib-users/.

D.3 Books

There are a lot of books on Python and there is no way I can provide reviews for all of them. The book by Mark Lutz, *Learning Python*, published by O'Reilly Media, provides a fairly comprehensive, if verbose, introduction for non-scientific Python.

For Pandas, the book by its originator, Wes McKinney, called *Python for Data Analysis* provides a thorough but terse treatment, and slanted toward applications in finance. There are a number of well-written tutorials on the web covering many aspects of Pandas, but it takes some digging to find ones that are useful.

The only text I have found that provides a good introduction to PyQt is *Rapid GUI Programming with Python and Qt* by Mark Summerfield. Unfortunately, it describes PyQt version 4, which was superseded by version 5 in 2013. Version 5 differs significantly from version 4, particularly in the coding of signals and slots, which makes version 5 incompatible with version 4. Nevertheless, Summerfield's text is useful even for learning PyQt5. A translator from PyQt version 4 to 5 can be found at https://github.com/rferrazz/pyqt4topyqt5.

Index

addition (+), 10
adjust space around plots, *see* `subplots_adjust`
animation, 287
    annotating animations, 292
    blitting, 298
    combining multiple animation instances, 308
    combining videos with animated functions, 306
    dynamic text, 293, 299
    fixed number of frames, 295
    function animation, 294
    function call, 298
    histogram, 307
    indeterminate number of frames, 300
    save to file, *see* FFmpeg
    sequence of images, 287
    static elements, 296
    until a condition is met, 300
anonymous functions, *see* lambda expressions
`append`, 95, 289, 292, 294
argument, *see* function
arithmetic, 10
    order of operations, 10
array (NumPy), 41
    attributes, 173
    Boolean masks, 47
    conditional (`where`), 160
    creating, 49
    differences with lists, 52
    instance variables, 57, 173
    masked, 112, 122
    mathematical operations, 43
    matrices, 49
        matrix operations, 51
    methods, 57, 173
    multidimensional, 49
        indexing, 51
    printing, 68
    slicing, 46
assignment operator (=), 13
attribute, *see* object

backend, 133
binary arithmetic operations, 10
blitting, *see also* animation
Boolean operators, 86

case sensitive
    commands, 5
    variable names, 15
class, 316
    attributes, 315
    defining, 318
    instance variables, 315
    methods, 315
code indentation, 88

comma after assigned variable, 297
comma after variable name, 297
comparisons, 86
complex, *see* numeric types
conditional statements, 82
- `where` function, `if-else` for NumPy arrays, 160
- `if`, 85
- `if-elif-else`, 82
- `if-else`, 84

configure subplots icon, 118
constructor (`__init__()` method), 319
curve fitting
- exponential using linear regression, 187
- linear, 175
  - with weighting (χ^2), 179
- nonlinear, 193
- power law using linear regression, 192

DataFrame, 242
- creating, 259
- `index` attribute, 262

dates and times, 251
datetime
- `replace`, 278

dictionaries, 53
differential equations, *see* ODEs
discrete (Fast) Fourier transforms, *see* FFT
division
- floor (`//`), 10
- normal (`/`), 10

docstring, 63
documentation

books, 361
help, 26
online, 359

`enumerate`, 91
event loop, 334
event-driven programming, 327
exception handling, 322
exponentiation (`**`), 10

FFmpeg, 291, 292
- installing, 343

FFT, 231
figure
- navigation toolbar, 100

figure size
- `figsize`, 121

floating point, *see* numeric types
function, 155
- anonymous functions, *see* lambda expressions
- argument
  - passing function names, 164
  - variable number (`*args`, `**kwargs`), 163
- derivative (`deriv`), 165
- keyword argument (`kwarg`), 74, 162
- namespace, 167
- passing function names and parameters, 164
- passing lists and arrays, 169
- passing mutable & immutable objects, 171
- passing mutable and

Index

immutable objects, 170
positional argument, 162
universal function (`ufunc`), 161
unnamed arguments
    `**kwargs`, 166
    `*args`, 164
user defined, 156

generator function, 300, 304
`glob`, 290
graphical user interface, *see* GUI
Greek letters in plots, *see* LaTeX in plots
GUI, 315, 328

imageopen `Image.open()`, 290
importing modules, 24
    matplotlib, 24
    NumPy, 20
    Pandas, 239
    SciPy, 194
    user-defined class, 317
`imshow()`, 290
`__init__()` method, *see* constructor
`input` function, 61, 63
    type casting, 63
installing Python, 4, 339
instance variable, *see* object
instantiate, 315
integer division, 12
integers, *see* numeric types
integrals, *see* numerical integration
interactive mode, `plt.ion()`, `plt.ioff()`, 100, 297
IPython, 4

magic commands, 6, 8
navigation commands, 6
system shell commands, 8
tab completion, 8
IPython pane, 4
iterable, 39
iterable sequence, 90

keyboard input, 61, *see also* `input` function
keyword argument, *see* function
`kwarg`, *see* function

lambda expressions, 171
    `set_printoptions`, 68
LaTeX in plots, 126
    fonts, 128
    Greek letters, 129
    math symbols, 130
    non-italicized text in math mode, 129
least squares fitting
    linear, 175
    with weighting (χ^2), 179
linear algebra, 212, 365
    eigenvalue problems, 214
        banded matrices, 216
        Hermitian matrices, 216
    matrix determinant, 212
    matrix inverse, 212
    systems of equations, 213
linear equations
    see , 217
list comprehensions, 94
lists, 35
    differences with NumPy arrays, 52
    multidimensional, 40
    slicing, 37

logarithmic plots, 116
    `set_xscale`, 119
    `set_yscale`, 119, 122
    log-log, 118
    semi-log, 116, 122
logical operators, 86
loops
    array operations, 93
    `for`, 87
        slow, 94
    `while`, 91
        infinite loop, 92

masked arrays, 112, 122
math text in plots, *see* LaTeX in plots
matplotlib, 20
    artist layer, 135
    backend, 132
        Qt5Agg, 133
    PyPlot scripting layer, 137
        backends, 137
        state machine, 138
    software layers, 132
matrix operations, 51
method, *see* object
module, 18
multiplication (*), 10

namespace, 21, 167
nonlinear equations, 217
    bisection, 220
    Brent method, 217
    Newton-Raphson, 220
    Ridder, 220
    systems, 221
numeric types, 10
    complex, 12
    floating point, 11
    integers, 10

numerical integration, 221
    single, 222
    double, 226
    double integrals, 226
    methods, 221
    polynomials, 224
    single integrals, 222
NumPy, 19, 22
    functions, 22

object, 173
    attributes, 57, 173
    instance variables, 56, 173, 174, 317
    methods, 55, 173, 174, 316, 318
object-oriented programming, 2, 55, 315
objects, 55
ODEs, 227
OOP, *see* object-oriented programming

Pandas, 20, 239
    `agg` method, 280
    `apply` method, 278
    `axis`, 278
    Boolean indexing, 258
    conditional indexing, 258, 264
    data from web, 261
    DataFrame, 256
        indexing, 257
        indexing with `iloc`, 257
        indexing with `loc`, 258
    dates and times, 251
    `dt.total_seconds`, 279
    `dtypes`, 260
    `groupby` method, 273
    `groupby` object, 274

Index 367

head method, 257
iloc method, 257
loc method, 257
pd.to_datetime, 278
plotting, 267
reading data, 240
　csv files, 240
　Excel files, 250
　keywords, 249
　text files, 247
selecting data, 264
Series, 253
sorting data, 263
statistical methods, 265
　table, 266
tail method, 257
time Series, 255
PEP 8, 28
plotting, 99
3D plots, 149
adjusting space around subplots, 118
bring figure pane to front, 125
color codes, 107
contour plots, 140
error bars, 108, 197
excluding points, 112
layering (order) of plot elements: zorder, 106
line and symbol types, 107
log-log, 118, 122
masked arrays, 112
meshgrid, 139
multiple axes, 125
OO interface, 118, 132
PyPlot, 102, 132
semi-log, 116, 122
set axis limits, 111, 123
streamline plots, 144
subplots, 113, 122
　grid, 122
　unequal grid, 198, 269
subplots, 122
tight_layout, 118
two x axes, 126
polar plot, 304
positional argument, *see* function
power (**), 10
print function, 18, 64
　formatted, 64
　suppress new line, 90
　with arrays, 68
program, 16
Pyflakes (syntax checker), 28, 340

random numbers, 209
　integers, 211
　normally distributed, 210
　uniformly distributed, 210
range function, 38
reading data from a file
　csv file, 71
　　Pandas, 240
　Excel file (Pandas), 250
　text file, 69
　　Pandas, 247
glob, 290
remainder (%), 10
reserved words, 15
routine, *see* program

save animation to movie file, *see* FFmpeg
SciPy, 19
script, *see* program
self parameter, 320
sort data

NumPy arrays, 57, 175
    using Pandas, 263
special functions (SciPy), 206, 208
    Airy, 206
    Bessel, 206
    error, 206
    gamma, 206
    Laguerre, 206
    Legendre, 206
    random numbers, 209
spinbox, 327
Spyder Window, 4
strings, 34
    concatenation (+), 34
    `split()`, 56
`subplots_adjust`, 118, 143
subtraction (−), 10
syntax checker (Pyflakes), 28, 340

tab completion, *see* IPython
TeX in plots, *see* LaTeX in plots
three-dimensional plots, 149

ticks
    color (parameters), 126
    manual labels, 124
    manual placement, 124, 151, 197
try–except, *see* exception handling
tuples, 35, 39
    multidimensional, 40

universal function (`ufunc`), 161
updating your Python software, 341

variables, 13
    legal names, 14
vectorized code, 161

wrapper, 1, 19, 133
writing data to a file
    csv file, 76
    text file, 73

`zip`, 59, 74, 77